BUILDING EXPERIMENT

BUILDING EXPERIMENTS

Testing Social Theory

David Willer
Henry A. Walker

Stanford Social Sciences
An imprint of Stanford University Press
Stanford, California

Stanford University Press
Stanford, California

Printed in the United States of America on acid-free, archival-quality paper

Library of Congress Cataloging-in-Publication Data

Willer, David.
Building experiments : testing social theory / David Willer and Henry A. Walker.
p. cm.
Includes bibliographical references and index.
ISBN 978-0-8047-5245-9 (cloth : alk. paper) — ISBN 978-0-8047-5246-6 (pbk. : alk. paper)
1. Sociology — Methodology. 2. Experimental design. I. Walker, Henry A., 1943–
II. Title.
HM514.W55 2007
301.072—dc22

2007001002

Typeset by Newgen–Austin in 10/14 Minion

Contents

Table and Figures

Table

Figures

Preface

MANY FORMS OF INVESTIGATION, FROM FIELD STUDY TO analytic survey, seek to emulate the laboratory experiment. Why is the lab experiment the methodological ideal? Lab research is more precise than other kinds because it is carried out under optimal conditions for observation and measurement. But that is not the reason it is the ideal. The experiment is the ideal because of its logical structure and because, unlike other forms of investigation, that logical structure can be fully realized in the laboratory. There is considerable utility in knowing the logic of experiments. Knowing that logic is essential to the budding experimentalist seeking guidance in research design. To know how any investigation departs from that ideal is useful to anyone who would design a research project or who would critically understand a project designed by another.

A key to understanding experimental design is that there is not a single logic of the laboratory experiment, but two logics. Both are covered in detail in this book. For the *theory-driven experiment,* the design is given by the theory. The purpose of this kind of experimentation is to test theory and, to that end, theory designs the experiments that test it. This kind of experiment is at least four centuries old. As we show, the logic of theory-driven experimentation crosses the sciences. It is the same in sociology as it is in physics. Having read this book, few, if any, readers would agree that fundamental differences between the sciences make experimentation a less plausible and less useful research strategy for sociology than for other sciences.

The second kind of logic, of the *empirically driven experiment,* is founded on the method of difference. The purpose of this kind of experimentation is to discover new phenomena, new relations in the world by constructing at least two circumstances as similar to each other as possible—but for a single difference. A discovery is made if that difference is followed by distinct outcomes. This kind of experimentation is also quite old and is used today under the same logic across the sciences. Compared to the theory-driven kind, the empirically driven experiment has the advantage that one need not await the development of a theory to begin investigation. It has the disadvantage that its results can never be as sure or as general as those of a theory-driven experiment. The two kinds of experimentation, individually or together, form a powerful methodology for the advancement of scientific knowledge.

We conceived this book as a way to introduce the logic, techniques, and procedures of laboratory experimentation to broader audiences. One such audience is those currently learning the methods of social research. This is a methods text, which, through the extensive use of examples, provides the understanding needed to conduct experiments. Although this is a methods text, it is also something more than a methods text.

To the active scholar nothing can be more important than knowing what a method of investigation can and cannot do. That broad array of active scholars is the second audience to which this book is addressed. The literature on the logic of experimentation, how experiments are designed and run, and the kind of knowledge that results is at best incomplete. Nowhere in the methodological literature will you find the logic of theory-driven experiments laid out, their design explained, and examples given of appropriate procedures. Only one of the two types of experimental research is recognized, yet one finds broad generalizations about experimentation that are simply false. For example, it is widely asserted that generalizing experimental results from the laboratory is difficult, perhaps impossible. Importantly, the results of theory-driven experiments are not generalized. Very different procedures, procedures that we explain, quite effectively bring their results to bear outside the lab.

Chapter 1 introduces the experiment in broad outline. In Chapter 2 we focus on how scientists carry out their work beginning with the questions that motivate research. We trace how a simple theory is stated and how an experiment is designed to test it. In Chapter 3 we analyze empirically driven experiments, reconstructing their logic as it is embodied in Mill's canons and in the statistical methods pioneered by Fisher. As a practical guide, we critically analyze a series of example experiments.

Are experiments in sociology different from experiments that have contributed to the explosive growth of knowledge in other sciences? Chapter 4 answers that question by showing that the logic of theory-driven experiments is identical across widely different sciences. Again a series of critically analyzed example experiments show the researcher how to design effective studies. Chapter 5 looks at the experimenter-subject relationship. We cover the ethics and humane treatment of research subjects, the experimental setting as a source of artifacts, as well as experimenter and subject bias. Here the reader is introduced to computer-mediated experimentation and its advantages insofar as minimizing artifact are concerned. In Chapter 6 we show how small, simple, and idealized social systems in the laboratory are connected to events in large, complex, and dirty systems outside it. We conclude by revisiting the logic of theory-driven experiments to show how other forms of investigation can take the same logical form and have comparable results.

We thank Kate Wahl, our editor at Stanford University Press, for her unswerving support of this project from proposal review through final editing. We also thank the National Science Foundation, which supported much of the research analyzed here. The first draft of the manuscript was written while Henry A. Walker was on sabbatical: thanks to the College of Social and Behavioral Sciences at the University of Arizona. We thank Barry Markovsky, Susan H. Roschke, Brent Simpson, Shane Thye, and students in the experimental methods course at the University of South Carolina as well as reviewers for Stanford Press for invaluable comments on the manuscript. Finally we express our gratitude to Patricia Powell Willer and Joyce A. Walker. Without Pat and Joyce's patience and support nothing could have been done.

David Willer, Columbia, SC

Henry A. Walker, Tucson, AZ

BUILDING EXPERIMENTS

1 WHAT IS AN EXPERIMENT?

THIS BOOK IS A GUIDE FOR DESIGNING AND EXECUTING social-science experiments. Like many of you, we were introduced to experimental design when we took undergraduate and graduate courses in statistical methods and inference. Unfortunately, the notes we took and the articles and books we read for those courses were of little use when we began designing our own experiments. Our colleagues offered advice and their critical commentary helped us improve our experimental designs. Yet for the most part, we learned how to build scientific experiments by trial and error. This book shares with you our hard-won understandings.

Our treatment of experimental design differs in approach and focus from others we have seen. Three ideas are essential to our approach: First, developing theoretical understandings of the relationships between phenomena is the primary objective of science—any science. Second, theories must be tested in the fires of research and experiments are the *best* tools for testing theories. Third, experimental results are important to theory development. Some experiments uncover weaknesses in existing theory and point to the need for revision. Other experiments validate or support theory. Their findings can encourage researchers to expand a theory's scope or to make it more precise. Finally, some experiments discover new phenomena that require theoretical explanation. These ideas suggest to us a continuous process. Theory development motivates and shapes the design and execution of experiments. The outcomes of experiments motivate theory revision and expansion and, sometimes, new theory.

Experiments: A Definition

We often ask students in our undergraduate courses to create hypotheses and carry out studies to test them. Some students create sociometric maps of their interpersonal relationships or take part in controlled laboratory exercises. Others observe patterns of status and influence at work or in public places. Still others use government documents to study social and demographic differences among various racial or ethnic groups. Sometimes their research reports describe the results of "experiments" that test "theories." These studies are often well conceived and well executed. However, the hypotheses are not theories and only a few of the studies are experiments.

Textbooks and dictionaries offer many definitions of an experiment. The definition we use is consistent with ideas employed by other working scientists (e.g., Lederman 1993).

> Experiment: An experiment is an inquiry for which the investigator controls the phenomena of interest and sets the conditions under which they are observed and measured.[1]

It is not the experiment's logical relationship to hypotheses or theories that distinguishes it from all other forms of investigation. Most if not all other forms of investigation try to emulate the experiment's logical connections to ideas and guiding hypotheses. Instead, the experiment is distinguished by the activity of the researcher who determines the conditions under which the investigation will take place. Wholly or in part, the researcher engaged in experimental research creates, builds, and controls the research setting. By contrast, researchers who use *nonexperimental* methods, including participant observation, case studies, and most surveys, conduct their research under conditions largely given by the society within which it is conducted.

The process of building and carrying out experiments begins when a researcher selects a phenomenon of interest and identifies conditions (or factors) thought to be important for understanding it. For some experiments, she devises hypotheses—statements about relationships between phenomena—and identifies *initial conditions.* By initial conditions we mean those conditions at the beginning of an experiment that establish the framework for the experiment. Often, the initial conditions persist over the life of an experiment ending only at its conclusion. During the experiment, the researcher *measures* structures and processes, tracks changes, and records *intermediate* and *end conditions.* Measures are created by assigning qualitative or quantitative values to conditions. They are used to produce data. Seen as a whole, an experiment begins when a researcher introduces initial conditions to participants. It continues through measurement of the structures and processes under study and ends when outcomes are measured.

The research process does not end with the final set of measurements. Data must be analyzed and interpreted, and the findings compared with the ideas that shaped the experimental design. A *good* experimental result will answer the questions that motivated the work while a *better* experiment will also raise new and exciting questions that can be answered only with new research. In that sense, experimental research is a never-ending journey and, at every stage of that journey, research scientists are concerned with increasing their experiment's power to test hypotheses, particularly if the hypotheses are derived from theory. Later we will distinguish hypotheses drawn from ideas devoid of theory and hypotheses derived from theories.

Experimentation and Social Science

To what uses can the experimental method be put in sociology and the social sciences? Some claim that experiments in sociology are limited to small group processes that are transitory and, perhaps, trivial. Nothing is further from the truth. Bureaucracies are the largest social structures built by humans and it has been known since Max Weber ([1918] 1968) that they function as they do because power is centralized. That is, hierarchies like the U.S. government and the Catholic Church work as they do because of power centralized in the hands of the president and pope respectively. Experiments discussed later in this book use exchange network theories of power to show how power centralization is produced by network structures like bureaucracies. In sociology, as in any science, the most important experiments test theory and are designed by theory. Only the power of theory and the limits of human imagination constrain the range of experimental research.

Conjoining theory and experiments allows large and complex social structures to be captured by theory and cut analytically into smaller and more tractable parts—parts that are subject to experimental investigation. Working in reverse, it is through theory that experimental results are combined into larger and larger wholes and thus brought to bear on large structures. Nevertheless, some sociological experiments focus on small group processes and, when they do, theory is as valuable there as when experiments are brought to bear on larger structures. There is no mystery about how theory and experiments are used together. Later chapters will examine, in detail, the connections between theory and experiments as we take you step by step through the design and execution of theory-driven experiments.

This book distills much of what we have learned about the design and execution of experiments and how theory and experimentation can work together as a powerful methodology. If this book leads you to consider experimentation for your next research project, helps you design successful experiments, or helps you to evaluate the designs and results of others' experiments, it will have contributed to progress in our science. If it can help you to avoid some of the hazards and pitfalls we have encountered, the paths you take as you build and carry out experiments will be smoother than those we have traveled. Let us begin our journey by looking at some example experiments.

Three Example Experiments

Our definition of an experiment and comments about the uses of experimentation in social science may offer little insight to those who have had limited contact with experiments or scientific research. What is the best way to deepen that insight? We take our lead from Einstein:

> If you want to find out anything from the theoretical physicists about the methods they use, I advise you to stick closely to one principle: don't listen to their words, fix your attention on their deeds. (Einstein 1954)

Here and in later chapters, we use an experiential approach. Abstract discussions of experimental design and procedures are supplemented with examples of *real* experiments that test *real* hypotheses and theories. Now we will describe three sociological experiments to show how the parts of the experiment—initial conditions, measurement procedures, and end conditions—are organized into an integrated whole. The examples show that experiments can be carried out far from laboratories and do not need sophisticated computing equipment. Once the three have been described, we use them to correct wrongheaded comparisons of experiments with other methods such as surveys and participant observation.

How Do Groups Organize?

In this section we focus on the laboratory investigations of Robert F. Bales (1950) and his associates at Harvard. Bales ran path-breaking studies that controlled the conditions of group formation and developed precise measurements of group members' interactions. Bales began studying group formation and other fundamental group processes in the late 1940s. The initial studies did not test formal hypotheses. Instead, they were designed to ask and answer several questions about newly created groups: Would they form differentiated structures under controlled conditions? If so, what kinds of structures would they create and why would they form them? The point of departure was the idea that people form groups to accomplish specific purposes. Bales presumed that individuals would establish (or have established for them) patterns of interaction, including hierarchical patterns of influence and status. So, how do collections of individuals that are not groups become groups?

Studying male undergraduates (usually sophomores) at Harvard, Bales and his assistants established several initial conditions. The research staff screened subjects to ensure that those in a given experiment did not know one another. The students sat around a table in a room equipped with one-way glass. An identifying number was placed in front of each subject so that research assistants could identify them. Experimenters gave the young men a discussion task and asked them to come to a unanimous decision about an issue they were given to discuss.

One discussion task is the case of Billy Budd. Based on a novel by Herman Melville, Billy Budd, a young seaman with a terrible stammer, is a victim of harassment, trickery, and a conspiracy to get him to lead a mutiny. Eventually, Budd strikes and kills an officer who has confronted him. Naval rules permit hanging or a lesser punishment for those who strike a superior. Participants in the experiment are asked to make a unanimous recommendation concerning Budd's punishment: Should he be hanged or given a less harsh punishment?

Controversial issues like the Billy Budd case typically evoke lively discussions, and the researchers' job is to observe patterns of interaction that develop during the discussion. How and what does a researcher observe during group interactions? What counts as interaction? Bales and his associates had to answer these questions before they could move ahead to formulate theory to address fundamental issues in group processes.[2]

Bales developed an elaborate scheme for classifying and coding interaction. The system used a series of numbers and symbols to record every verbal or nonverbal act (e.g., who made a comment and to whom it was directed). Eventually, Bales devised rules for classifying behavior into 12 categories. Those categories were grouped into three general classes (1) *positive social-emotional behavior* like praising another's actions, (2) *negative social-*

emotional behavior such as expressing hostility toward another member, and (3) *task behavior* like asking for or giving directions about the task.

Initially, Bales and his research assistants used a moving scroll to record acts in real time, a recording method that allowed interaction processes to be reviewed systematically. A typical record showed a sequence of task behaviors: Subject 1 asked Subject 3 for help that Subject 3 subsequently gave to Subject 1. As technology evolved, researchers began using audiotapes, then videotapes, and, eventually, computer-assisted videotaped recordings to observe interaction.

Bales' research produced many important findings. Among them was the discovery of a stable structure of interaction as measured by the distribution of participation across group members (i.e., the time that each member spent talking). Most groups developed clear patterns of interaction after only 15 or 20 minutes of discussion. Consider that each member of a three-man group would talk 33 percent of the time if members participated equally. In Bales' groups, one person usually did the lion's share of talking and one talked very little. For the average three-man group, the most active member talked about 44 percent of the time while the least active member did only 23 percent of the talking. In eight-man groups, if talking were evenly distributed across members, each would talk 12.5 percent of the time. However, data show that the most active talked 40 percent of the time and the least active only 3 percent of the time. The patterns were remarkably stable. Typically, when groups were called back to the lab for a second session, the same people were ordered in the same way.

Bales' early studies are unusual, even unique among the experiments discussed in this book in that they did not test hypotheses. Nevertheless, they were experiments because the phenomena of interest (e.g., group composition and size) and the conditions under which observations were carried out were established and actively controlled by the investigators. Furthermore, these were among the first sociological experiments to develop elaborate measurement procedures—a pioneering system for coding interaction. They also uncovered a puzzling finding. Some groups seemed to form structures almost instantaneously as if they had formed them before the study began. How and when those structures formed became clues inspiring the development of Status Characteristic Theory (SCT), a theory of quite substantial explanatory power that links status and influence processes. We will return to Bales' research when we discuss experiments that test hypotheses drawn from Status Characteristic Theory.

A Study of Centrality and Benefits in Exchange

Alex Bavelas began the study of connections between the shapes of communication structures and task outcomes like productivity, efficiency, and so on in the late 1940s. He studied highly centralized structures like the "star" or "wheel" structure in which one position is connected to every other position; and the other positions are connected only to the central one. Less centralized structures included the circle—in which every position is connected to two other positions such that they form a closed chain or circle. Among his early findings, Bavelas (1950) reported that groups with centralized structures generally had higher productivity, solved tasks more quickly, and made fewer task errors.

Diagrams of Bavelas' communication structures look very much like diagrams of leadership structures commonly found in groups and formal organizations. Some are highly centralized (e.g., authoritarian structures) while others are more egalitarian (Lewin, Lippitt, and White 1939). Graph theory and network analysis (Flament 1962; Harary, Norman, and Cartwright 1965), were introduced in the mid-1950s making it possible to use quantitative techniques to analyze social structures—represented as graphic diagrams—and positions of those structures. Those developments opened entirely new areas of sociological inquiry. For example, consider exchange networks in which two or more people can trade—or exchange—goods and services. Does a person's position in an exchange network affect the benefits she can gain in exchange? If so, how?

Two exchange networks are shown in Figure 1.1. The first, a triangle structure (Figure 1.1a) shows three people, Alice, Brenda, and Carol. The lines show

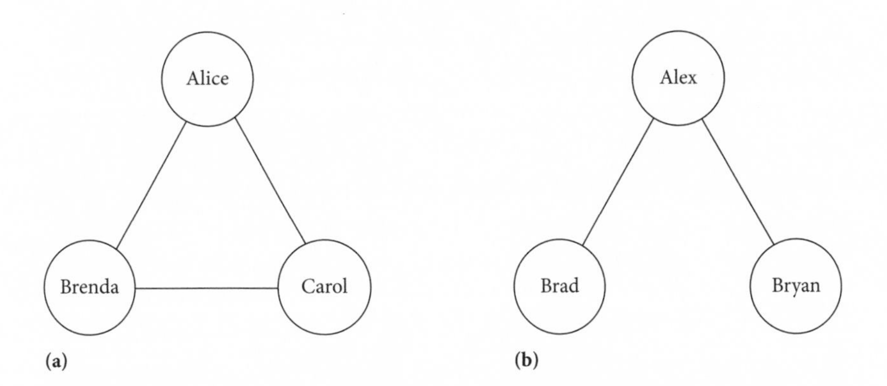

FIGURE 1.1. Decentralized and Centralized Exchange Structures

which pairs can complete an exchange. The network has three members and each can complete a trade with every other member. Figure 1.1b is very different. Alex, Brad, and Bryan are also members of a trading network but no line connects Brad and Bryan. They cannot make a trade, although each can exchange with Alex. Figure 1.1b is centralized; Alex's position is different from the other two and can be distinguished even when the names are removed. By contrast, the three positions of Figure 1.1a are identical and cannot be distinguished once the names are removed.

Is it possible that central positions like Alex's are advantaged in trading networks? Do they get more benefits than less central positions? We restate the question as Working Hypothesis 1:

> H_W1. The payoff that a position gains through exchange in each relation varies directly with the centrality of its location in an exchange structure.

Let centrality be measured by the number of relations connected at a position. Therefore, Alex is more central than his two partners, whereas Alice, Brenda, and Carol are equally central. The hypothesis implies that Alex will gain more than Brad in their exchanges and more than Bryan in their exchanges, but that Alice, Brenda, and Carol will gain equal payoffs in their exchanges. The experiment below was designed (1) to learn whether being centrally located in an exchange network is advantageous and (2) to measure the size of any advantage.

The experiment is concerned only with the effects of centralization so it is reasonable for an experimenter to create structures that differ only in terms of centralization as in Figures 1.1a and 1.1b. Experimental control can ensure that they are otherwise as similar as possible. To accomplish this, each pair of trading partners is connected by the same kind of standardized relation. Acting in the standardized relation is like dividing a pie. If one person's piece is larger, the other's piece will be smaller. Both must agree on a division for either to receive any pie at all. People in this relation are assumed to be self-interested in that each wants to receive as much pie as possible and both want to avoid disagreement when neither receives a pie payoff. Their interests are both opposed and complementary—just as in naturally occurring exchanges. Game theorists call these kinds of relations "mixed-motive" games. People compete for benefits (pieces of pie) but must cooperate to get any benefits.

Our example experiment builds a standard relation, called a *resource-pool relation*. Not a pie, but a pool of valued resources is placed between a pair of subjects. Call them A and B. If and only if A and B agree on a division, will they each receive the agreed on part of the pool. The pool in our prototype relation contains 10 resources and, if both agree that A should get 7 resources, then B will gain $10 - 7 = 3$ resources. If instead, both agree that B should receive 6 resources, then A will receive $10 - 6 = 4$ resources. The resource-pool relation is *like* an exchange relation in that (1) A and B both benefit when they agree but (2) neither benefits if they fail to agree.

The experimenter establishes four initial conditions. The resource-pool relation is the first initial condition. The second initial condition establishes the subjects' preferences for more rather than fewer resources by paying each according to the number of resources gained. For example, each might earn $0.50 for each resource received. For the third initial condition, the researcher builds two structures like those in Figure 1.1. Using the same prototypical relations for both designs makes it easy for the researcher to compare results observed in the two situations. The fourth condition specifies that, for a given period, each position may exchange in each of its relations. Thus, Alice, Brenda, and Carol each can exchange twice, once with each partner—as does Alex. Because Brad and Bryan are connected only to Alex, each can exchange only once.

The experimental design must be given a concrete setting, and researchers have used two, one low tech and one high tech. Both settings can run the triangle and centralized designs in Figure 1.1.[3] In the low-tech setting, subjects sit in chairs facing each other and negotiate with resources represented by a stack of counters. Poker chips make fine counters. Barriers separate subjects not connected by exchange relations. In the high-tech setting, specialized software allows subjects seated at PCs in separate rooms to see the centralized or the triangle structure on their computer screens. The screen shows the subject where she is located in the exchange structure, the positions to which she and every other position is connected, and the value of the resource pool relations she shares with others. Using mouse control each subject can send offers of resource divisions and make exchanges. All subjects' screens show every offer and every completed exchange.

These structures can be studied as *repeated games*. That is, each group of subjects has several opportunities, called rounds, within which to negotiate and complete exchanges. Each round begins with fresh resources placed between each pair of subjects. Then subjects negotiate, reach agreements, and divide resources. A round ends when all possible divisions have been completed or when a time limit, usually 3 minutes, has expired. The experiment has several sessions with a new set of subjects recruited for each session. The measurement of outcomes—the end conditions of the experiment—completes both experimental designs. The outcomes are the resource-pool divisions agreed to by the subjects. An observer records those resource divisions in the face-to-face situation. In the PC system, all offers and exchanges are automatically recorded and stored in the memory of a central computer and can be downloaded later.

The results from experiments like these are clear: occupying the central position in a social structure is not advantageous relative to other positions. Nor is it disadvantageous. Observations of the triangle with identical positions found approximately 5 − 5 divisions for each exchange. Researchers also found that mean resource divisions in the centralized structure were approximately 5 − 5. Furthermore, results are very similar for the low- and high-tech settings. It follows from these results that Working Hypothesis 1 is wrong. Central location is not advantageous in exchange. The results also show that the widely held belief (call it the commonsense view) that centrality is beneficial either is wholly wrong or must be qualified by these results. Nevertheless, as we will show later, social theory has uncovered structural conditions that advantage central positions. Just as important, theory also identifies structural conditions that disadvantage central positions.

Looking back, the initial conditions of the example experiment include the prototype resource-pool relation. Three resource-pool relations form the triad and two resource-pool relations form the centralized structure. Because subjects were paid according to resources gained in exchange, the sparse environment of the experiment makes it reasonable for experimenters to infer that only the payoffs to be gained from those relations motivated the subjects' actions. The resource-pool relations were appropriate to test the hypothesis because payoffs for those relations are identical to those for many exchange relations.

A Field Experiment

Field experiments are like lab experiments in that the experimenter controls key elements of the research setting. Field experiments are unlike lab experiments in that, being carried out in the field, some study conditions are not under the experimenter's control but are, instead, given by the society in which the study is conducted. We describe a field experiment carried out by Berk, Lenihan, and Rossi (1980) that focused on an important practical issue.

Criminologists have documented a strong correlation between poverty and criminal behavior. Official statistics also show that an overwhelming majority of prison inmates return to prison in the period immediately following their release. The rate of return (called the *recidivism rate*) creates serious problems for criminal justice systems. Prison populations rise dramatically when many repeat offenders are added to inmates serving time for their first conviction. Berk et al. designed the Transitional Aid Research Project (TARP) to explore the effects that institutional support, including monetary payments, have on recidivism rates. We state the principal hypothesis as Working Hypothesis 2.

> H_W2. Prison inmates who receive institutional support following their release are less likely to return to prison than those who do not receive institutional support.

As did researchers in the two earlier experiments, Berk, Lenihan, and Rossi established several initial conditions. They drew a sample of female and male inmates released from prisons in Texas and Georgia. With cooperation from the Department of Labor, some of those released were given help with job placement and awarded weekly stipends for either 26 or 13 weeks. If the former inmates found jobs during the study period, their stipends were reduced either $1 for every dollar earned (100 percent tax) or $1 for every $4 earned (25 percent tax). Another group of inmates received job placement assistance but no stipends. The researchers also studied two other groups. One group had periodic interviews with research staff but received neither stipends nor assistance with job placement. A final group was monitored by research staff, but the staff was not actively involved with them. In summary, the study followed six groups of inmates who:

- Received job placement assistance and a stipend for 26 weeks subject to 100 percent tax against wages earned.[4]
- Received job placement assistance and a stipend for 13 weeks subject to 100 percent tax against wages earned.
- Received job placement assistance and a stipend for 13 weeks subject to 25 percent tax against wages earned.
- Received job placement assistance.
- Were interviewed at periodic intervals by research staff.
- Were monitored but not directly engaged by research staff.

The TARP findings generated much discussion among academic researchers and policy makers but the results were clear. Payments to released prisoners did not substantially reduce recidivism rates for former inmates in either Georgia or Texas. Within one year of release, approximately one-half

(50 percent) of male prisoners had been returned to prison for new offenses. The numbers of women released into the study were too few to allow researchers to complete detailed statistical analysis for female offenders. However, the data for male offenders released from Georgia and Texas prisons were nearly identical. The data fail to support Working Hypothesis 2.

A Brief History of Two Kinds of Experiments

Our three example experiments show the range of social science experiments. Bales' research focuses on microsociological processes, the centrality experiments have implications for small and large structures (e.g., bureaucracies), and findings from the TARP study have implications for whole societies. Beyond these substantive observations lies a distinction central to this book: that there are two kinds of experiments in science—*empirically driven* and *theory-driven* experiments. Distinguishing the two types is important. Empirically driven experiments *explore or discover phenomena* while the theory-driven experiment *tests theories* (Hempel 1966). All three example experiments are empirically driven. In Chapter 2 a theory-driven design is presented.

Theory-driven experimentation, if traced to Archimedes, is more than 22 centuries old. Archimedes, writing in 230 BC, put forward his laws of levers and it seems that they were systematically tested ([230 BC] 1897). Whether Archimedes did that testing or not, it is certain that the *modern* tradition of theory-driven experimentation coupled with mathematical analysis can be traced to Galileo's tests of his theory of falling bodies. Though the Catholic Church delayed the publication of Galileo's work until 1636, it was completed in 1607. Therefore, the modern tradition of theory-driven experiments is now four centuries old (Galilei [1636] 1954). Both Archimedes' and Galileo's experiments are analyzed in Chapter 4.

The interplay between theory-driven experiments and theory development that began with Galileo continues across the sciences to this day. Having encountered Descartes' analytic geometry by 1664, Newton published *Principia Mathematica* ([1686] 1966), which refined and extended Galilean mechanics. Since then, first in physics then in chemistry and more recently in biology, the stream of work linking experiments and theory development has grown and grown again. Though mainstream methodologies in sociology owe little or nothing to these developments, as we show in Chapter 4, there is theory development and research in sociology that is identical to the great stream of work in the physical sciences.

By the middle of the nineteenth century, modern scientific experimentation was well established. It was so well established that there was every reason to suppose that it would be extended to the newly emerging science of sociology. However, that extension has not been realized. J. S. Mill's influential *A System of Logic* ([1843] 1967) claimed to explain how science works. However, it said nothing about the relations between theory and research and how the two interact to advance knowledge. Instead, Mill promoted experiments as the best tool for discovering empirical regularities. Today we recognize empiricism as the view that theories and laws are statements that describe regularities. Empiricism owes much to Mill who fell into the error of believing that the purpose of science was to find invariable regularities.[5] Mill and other empiricists and statisticians who followed him, most notably R. A. Fisher, institutionalized the methods subsequently adopted by most sociologists.

Now here is the irony. Over the 160 years since Mill, empiricist methods, though grounded in a misunderstanding of science, have been refined into powerful research tools. These research tools cannot produce theory and cannot advance sociology as a theoretic science. Nevertheless, the methods advanced by Mill and subsequently refined by Fisher have been fruitfully used to discover relative regularities, not just in sociology, but in psychological social psychology, medicine, agronomy, and many other fields. For any field where theory is little advanced, there is no better way to learn about the world than by Mill's and Fisher's experimental methods. We discuss in Chapter 3 the design of empirically driven experiments and show how they are built following the logic of Mill's and Fisher's designs. On one point we can agree with Mill: His kind of experiment is an excellent method of discovery. On the other hand, the theory-driven experiment is the best method for testing theory and advancing scientific understanding.

Comparing Experiments and Nonexperimental Studies

Experiments are the best method for testing theory because they have many advantages compared with other research methods: (1) They give researchers control of some or all conditions under which phenomena are observed, (2) they permit more precise measurement of factors and processes of interest, (3) they permit researchers to create and reproduce the initial conditions necessary to test hypotheses drawn from a theory, which (4) has the effect of improving the fit between theory, hypothesis, and research findings. This combination of advantages makes for powerful tests. As a result, an experiment that generates positive findings lends strong support to hypotheses. Similarly, experiments that produce negative findings offer equally strong grounds for rejecting hypotheses. For these reasons experiments offer better tests of hypotheses than other data gathering techniques. Given these and other advantages of experimental research, objective observers ask why very little research in sociology uses experimental methods.

Our answer has three parts: First, few sociologists are taught the purposes of experiments or how to design and conduct them. Second, few sociologists have been motivated to learn experimental techniques because it is widely believed that experiments cannot produce useful sociological knowledge. Third, some experiments can raise sticky ethical issues slowing the human subjects approval process that must be completed before research is begun.

This book is intended to resolve the first issue by explicating, in detail, the logic of empirically driven and theory-driven experiments and the principles of good design for both types. It also discusses sticky ethical issues and how to resolve them. Those issues are dealt with in later chapters. Here we confront the belief that the subject matter of sociology is not suitable for experimental research.

Those who are skeptical of the utility of sociological experiments offer several reasons. First, many point out that the controlled conditions of experiments are artificial. Consequently, there is no reason to suppose that people face similar conditions outside the experimental setting. Second, experimental subjects know that they are in an experiment and under observation. By contrast, people who are studied as they go about their routine activities (as in some case studies and some forms of participant observation) need not be aware that they are objects of investigation. It is plausible that just knowing that one is under scrutiny has substantial but unknown effects on behavior. Finally, experimental subjects are typically drawn from ill-defined populations by sampling methods that are either unknown or unclear.[6] Thus generalizing from experimental subjects to a larger population, a common practice among survey researchers, is impossible.

We can easily show that such reasoning is misguided by considering parallel arguments that survey research cannot produce useful knowledge—even though sociologists and other social scientists are undoubtedly the most sophisticated users of survey methods. Most surveys are carried out under conditions already present in the society—conditions that are not under the investigator's control. Lacking experimental control, findings rarely strongly support hypotheses or result in their unequivocal rejection. Precise measurement may be difficult, if not impossible, for research carried out in field settings. Generalization is limited to induction from a sample to a population, but every population is bounded by the particular conditions of a time that lies in the past. Because the investigator can generalize only to the past, explanation of current events and prediction of future events are impossible. Because general, universal knowledge cannot be attained, knowledge gained from surveys has little or no explanatory or predictive power.

Both critiques are misguided. The critique of the experiment amounts to the claim that the experimental method is faulty because it produces a poorly designed survey. The critique of the survey is similarly faulty. It asserts that the survey method is faulty because it produces a poorly designed experiment. In fact, the two methods are logically distinct, have distinct purposes, and neither should be judged by the standards applied to evaluate the other. The same is true of other methods like the case study and historical-comparative research. Each has its own strengths and weaknesses and should not be judged by the standards of experimental or survey research. When one method is judged to be the best, it is always best for some questions while for other questions it is less effective than other methods, or even wholly ineffective for it cannot be used at all. For example, the experimental method is the best for theory testing, but cannot be used to find prevailing opinions in a population.

There is no one best research method. For example, to learn about street-corner society today, what is needed is a case study, perhaps a series of case studies, undoubtedly carried out by participant observation. No survey, experiment, or historical investigation can gather the desired information. Looking back at the experiment on exchange and centrality, it is conceivable that field studies could investigate the relationship between payoffs and exchange centrality, but not well. It is unclear how a researcher would go about drawing a sample of centralized exchange structures and fully connected triangles that are similar enough to compare rigorously. Furthermore, the experimental structures described above were built in complete isolation. By contrast, no test has yet been devised to determine whether a structure studied in the field is or is not affected by other structures nearby.

Because they have been developed for different purposes, the research methods of sociology rarely compete, but often they can complement each other. A survey could ask a national sample of respondents the question: "Does being central in an exchange structure convey an advantage?" In analyzing responses, it does not matter that few if any of those responding are qualified to give answers. At issue is not knowledge but opinion. A survey can be used to find out whether the *belief* that being central is advantageous is widespread. Whether the belief is or is not widespread could well be significant to sociologists, economists, and consumer advocates. Consider the often-documented finding that members of particular ethnic minority groups pay more for automobiles (an exchange outcome) than members of the majority. Perhaps they pay more because more of them hold incorrect beliefs about the advantages that auto dealers have. If a survey finds that incorrect beliefs are widespread, experiments could investigate the impact of those beliefs on exchange.

The experimental method could not begin to address questions about opinions held by members of any population, but it can successfully investigate important social phenomena. How do some people exercise power over others? How is power created and how should it be measured? Can power be countervailed? Under what conditions do people conform to the expressed opinions of others? Under what conditions do they not conform? Does influence vary with status such that those of high status are more influential than those of low status? Can influence be converted to power? Why are some people willing to follow authority? Why do some resist? These are the kinds of research questions on which experiments can be brought to bear and they are the kinds for which experimentation is undoubtedly the best method. Furthermore, they are research questions that experiments have recently investigated, and social scientists have used the results of those experiments for explanation and prediction outside the laboratory.

Experiments are important tools for discovering phenomena and for testing theory. The relationship between experiment and theory is central to this book. Having introduced the experiment here, we now turn to theory and its relation to the experiment.

2 THEORY AND THE SCIENTIFIC METHOD

SCIENCE HAS AS ITS PRIMARY OBJECTIVES DESCRIPTION, CLASsification, and explanation of phenomena. Science begins with questions, and it must develop and test theories to answer those questions if it is to advance. This chapter begins by introducing a three-category scheme for classifying scientific questions. Answers to the three types of questions have different scientific purposes but researchers can use all three as part of a comprehensive strategy of theory development. Next, we give our definition of theory and use an example to show how ordinary language statements can be expressed formally. We describe several forms that theory can take and identify criteria for evaluating theory. Finally, we show how questions, theory, and experimental tests of theory are combined to create research programs that lead to theory growth.

From Questions to Answers

Every scientific investigation begins with a question but, as the three experiments we described in Chapter 1 show, the questions can be very different. The substantive differences between questions are less important than the forms the questions take and the kinds of answers they generate. We return to those experiments—not to restate the obvious—but to discuss important distinctions between the questions that prompted them.

Walker (2002) identifies three broad classes or types of scientific questions. The first class of questions asks about the existence and character of some phenomenon. A second type asks about relationships between phenomena, and the last group consists of questions that ask why two (or more) phenomena are related.

The first type of question that scientists ask is a sophisticated version of the questions that young children ask: What is this thing, *y*? How is a particular *y* similar or different from another *y*? Questions like these are the point of departure for all scientific inquiry. These Type I questions (Walker 2002) motivate researchers to make systematic and detailed observations of a phenomenon. Such intense scrutiny encourages development of rigorous descriptions of phenomena and systems for classifying them. The early experiments in the Bales (1950) research program discussed in Chapter 1 investigated Type I questions. Bales answered many of them by identifying a variety of verbal and nonverbal behaviors and by developing a general system for classifying them.

The second group of questions (Type II questions) asks: What factors are related to the occurrence of an event *y*, or members of an event class *Y* (i.e., the set of all *y*-like events)?[1] Because they focus on relationships—second order phenomena—Type II questions call for more complex descriptions than those generated by Type I questions. Once they had a reliable system for classifying behavior, Bales and his associates could identify patterned relations and covariation between types of behaviors. For example, Bales identified an association between group members' rank and time spent talking as well as how other members ranked the quality of their ideas. Answers to Type II questions can also motivate researchers to raise additional questions that ask for explanations. Some questions that ask for explanations are variants of Type II questions. They lead researchers to develop empirical, or statistical, explanations. Only one type of question motivates researchers to develop theoretical explanations—call them Type III questions. We will take up Type III questions after we discuss empirical explanations.

Empirical explanations answer questions about how one or more concrete events are related to (or "explain") a concrete event or variation in concrete events. They are understandings that focus on concrete, time-specific events or factors (*x*s or *X*s) that lead to (or "cause") the phenomenon under study (*y* or *Y*).[2] Social scientists are proficient at devising *qualitative* and *quantitative* empirical explanations.

Qualitative empirical explanations focus on explaining a unique event, *y*. For example, Weber ([1904] 1958) tried to explain the effect of ascetic Protestantism on the rise of capitalism in the West, itself a unique occurrence. In the aftermath of the events of September 11, 2001, policy analysts focused on each attack launched against American targets. Their primary objective was to identify the chain of concrete events (*x*s) that led to each attack. Analysts learned that (1) many attackers were Saudi nationals, who (2) entered the country legally and "disappeared" in American society, and (3) paid cash for one-way tickets, and so forth. One objective of this analysis was to find out whether government or law enforcement officers could have intervened at some particular *x* (or *x*s) to thwart the attacks.

Quantitative empirical explanations focus on explaining an event class, *Y*. More accurately, researchers explain variation in the phenomena that make up such event classes by identifying covariation with phenomena in other event classes, *X*s. As an example, Lichter, McLaughlin, and Ribar (1997) were interested in why some women were sole heads of families and others were not. Their analysis of county-level data shows that 13 variables (*X*s) are significantly correlated with the incidence of female-headed households (the dependent variable). The variables include women's earnings, population, the male-female sex ratio, and so on. Their initial model explained 83 percent of the variation in *Y*, female-headed households ($R^2 = 0.834$).[3] Analysts who use multivariate techniques to explain variation in a dependent variable are engaged in quantitative empirical explanation. We will discuss experiments that are based on statistical designs in Chapter 3.

Returning to experiments discussed in the first chapter, the study of positional centrality and the TARP study of economic support and recidivism respond to Type II questions. Each involves speculation about the causal connections between phenomena. People studying communication networks had observed that a position's centrality was related to its control of resources. On that basis they speculated that centrality was also related to the distribution of payoffs in exchange networks. Similarly, criminologists and social policy researchers knew that poverty and criminal behavior were related. The TARP study was designed to find out whether improving the financial resources of released prisoners would reduce recidivism.

Type III questions that motivate theory development offer a sharp contrast to Type II questions that find empirical explanations. Questions motivating theory development focus on relationships between theoretical concepts like X_C and Y_C (e.g., status rankings and influence), and can only be answered by devising a theory that includes the general mechanisms that account for the relationship. The history of many questions that motivate theory development includes answers to Type I (what is *y*) and Type II questions (which *X*s are related to *Y*). None of the experiments described in Chapter 1 responded to Type III questions about relations between theoretical concepts nor were they designed as tests of theory.

Because theory development is central to scientific progress and experiments are the best method for testing theory, the relation between theory and experiment is central to this book. Those new to theory-driven experiments may ask, "Where do I find a theory for my experiment?" We suggest two sources. First, a researcher can work with existing theories. If a researcher takes this path, the first task is to explore inferences that have not been investigated, or make new inferences from the theory. Second, a researcher can create new theory. Those who take this path may find themselves moving though all three types of questions. In either case, success will depend on an understanding of theory and how to use it. Therefore, we now turn to answering the question: What is theory?

Theory: A Definition and an Example

Theories explain relationships between phenomena.[4] The definition that follows uses ideas from one of our favorite theory construction texts (Cohen 1989:178).

> Theory. A theory is a set of interrelated, universal statements to which a set of rules or procedures can be applied to create new statements.

We unpack the definition by emphasizing that theories *must* contain interrelated statements. Interrelated statements are necessary to permit derivation of new statements and to explain the relationships under study. Now we describe a theory that answers the question: Why are positional centrality and payoffs positively related in exchange networks? Notice the form the question takes: Why is one phenomenon (positional centrality) related in a particular way (positively) to a second phenomenon (payoffs gained)? Call the following three arguments or theoretical statements (TSs), the Theory of Positional Centrality:

> TS1. **Centrality** of positions in exchange networks is *negatively* related with their **dependence** on other positions. Less central positions are more dependent.

TS2. **Dependence** of positions in exchange networks is *negatively* related with their **power**.

TS3. **Power** of positions in exchange networks is positively related with **payoffs**.

The three statements are interrelated because each statement shares a term with at least one other statement in the set. (See the terms printed in bold type.)

Theoretical statements must also be universal. They cannot refer to specific times, places, or things. None of the key terms, the theoretical constructs given in bold, in the theory above refer to particular times or places. In fact, general discussions of theory often replace natural-language theories, like the theory of positional centrality given above, with purely symbolic language like the following revised arguments:

TS1r. $A_C R_1 B_C$

TS2r. $B_C R_1 C_C$

TS3r. $C_C R_2 D_C$

We interpret statement 1 by writing, "A_C stands in a particular relation, R_1, to B_C." For statement 2, we write that "B_C stands in relation, R_1, to C_C," and similarly for statement 3. The terms in statements 1–3 are universal in the algebraic sense. A_C, B_C, C_C, and D_C together with R, the symbol we use for relation, are theoretical constructs that can represent any phenomena or relations. We have added subscripts to the Rs to show that the relation described in one statement can be distinct from the R in another statement in the same theory. As an exercise, convince yourself that the theoretical constructs and relations "centrality," "dependence," "positive relation," and so forth can be substituted for the symbols in statements TS1r–3r.

The requirements that theoretical statements are interrelated and restricted to universal terms have important implications for theory testing. Interrelated arguments permit scientists to create entirely new statements—a quality that distinguishes theories from isolated hypotheses. For example, rules of logic can be applied to the theory above to infer the statement we label Derivation 1 (D1). Note that Derivation 1 can also be stated in purely symbolic terms as, $A_C R_x D_C$.

D1. The **centrality** of positions is positively related to their **power**. (From TS1 and TS2.)

Two more inferences are given immediately below and we label them Derivations 2 and 3. As before, the derivations follow logically from statements 1–3 above.

D2. **Centrality** of positions in exchange networks is positively related with **payoffs**. (From TSs 1, 2, and 3.)

D3. A position's **dependence** is *negatively* related with **payoffs**. (From TS2 and TS3.)

With the Theory of Positional Centrality in hand, it is time to consider how it can be tested. In testing it we will create conditions under which observations can evaluate one or more of the statements of the theory. That researchers use observations to evaluate the worth of universals in theories seems to present a paradox because observations *always* refer to specific times, places, and events. How can scientists bring theories that do not refer to specific times, places, or things into contact with specific observations? The paradox is resolved because theories have the capacity to generate new statements.

Theories can be tested after new statements that link the universal and abstract to the particular and observable are generated. Such statements are called *linking statements* (LS below) because they link two terms, one abstract and theoretical and the other concrete and empirical (Cohen 1989:82). As below, the empirical term designates a method of measurement. We use the Theory of Positional Centrality and the experiment described in Chapter 1 to show how linking is done.

LS1. The number of relations connected to a position in an exchange network measures its centrality (initial condition).

LS2. The number of resources (poker chips, counters, etc.) received in a resource-pool division is a measure of payoffs gained in exchange (end condition).

Linking statements like LS1 and LS2 permit an investigator to substitute concrete variables for theoretical concepts and to derive new statements written in concrete observational terms. Derivations that contain observational terms (variables) are often called hypotheses to distinguish them from derived statements that contain only theoretical constructs. Hypothesis 1 (H1) is an example of a new statement that can be created by using the initial and outcome condition statements to interpret the Theory of Positional Centrality. Notice that H1 restates D2 above in concrete language. H1 is concerned with the same relation as the first working hypothesis in Chapter 1, H_W1.

H1. The number of relations connected to a position is *positively* related to the number of poker chips it receives in a resource-pool division.

Returning to the experiments we described in Chapter 1, Alice, Brenda, and Carol have the same number of connections, 2, and gained equal numbers of resources in each relation (approximately 5). For the centralized structure, Alex had 2 connections, more than Brad or Bryan who had 1. Yet the three people in the centralized network also gained equal numbers of resources in each relation (approximately 5).

Our discussion of theory and its empirical interpretation began with an example theory, the Theory of Positional Centrality that we stated as a series of three related universal statements. We have shown how the relations of the three statements allow us to derive new statements. We have also shown how universal statements can be given an empirical interpretation to bring them into contact with observations drawn from experiments. Looking back to the experiment discussed in Chapter 1, we can now conclude that the data from the centralized structure are inconsistent with H1. That is, H1 is false.[5] Because it is false, logic also requires a researcher to treat its universal equivalent, Derivation 2, as false. In fact, the data falsify the Theory of Positional Centrality.

Nevertheless, we are not disappointed with this theory's falsification because long experience has taught us not to confuse negative results with a failure to advance knowledge. Research programs that develop theories will have dead ends, many of which are marked by falsifications. Frequently those dead ends are learning experiences. Now we know that being central does not itself produce power differences. We also know, by looking outside the lab, that people in central positions are frequently very powerful. Because our knowledge of power in networks has been advanced, we now know that, to explain the power of those in central positions, we must look beyond centrality to discover the conditions that produce power. We will present some of those discoveries in later chapters.

Some Criteria for Evaluating Theory

Several criteria are useful for evaluating theories. They include (1) logical consistency, (2) breadth of scope, and (3) empirical support. A theory that includes or implies the statements "centrality is negatively related to dependence" and "centrality is positively related to dependence" is not logically consistent. The two statements are contradictory. Sometimes inconsistencies can be found by careful examination of the internal structure of a theory. More frequently, inconsistencies are found when the theory is applied and its predictions are logically impossible. In either case, when a theory is found to contain or produce contradictions, it is the obligation of the theory's authors to eliminate them—if they can. Until they do, the theory is useless.

Theories are normally applied within their designated *scope conditions.* Scope conditions are always stated in universal, not particular, terms. For example, theoretical economics—what economists call neoclassical microeconomics—is scope-limited to market economies. The term "market economy" is a universal term because it does not refer to specific times, places, or economic arrangements. Ancient China and Rome (Weber [1918] 1968:164) perhaps contained market economies, most contemporary societies contain market economies, and many that might appear in the future will undoubtedly have market economies. The statements, "all economies today" or "twentieth-century U.S. economy," cannot define the scope of theoretical economics because their terms are not universal. They refer to particular economies and particular times and places.

More generally, scope conditions define two domains, one where the theory applies and one—often broader—where the theory does not apply. Though theories are universal, there are general conditions under which they do not apply and, when properly drawn, scope conditions will rule out those applications. Thus market economy as a scope condition excludes application of neoclassical microeconomics to communal societies such as bands and tribes. This exclusion is important. Supply and demand may vary in economies without markets but neither affects prices because there are no prices. Any attempt to apply neoclassical microeconomics to such societies would be empty of implications. Moreover, the failure to find supply and demand effects on prices would not falsify the theory because nonmarket societies lie outside its scope.

A third consideration is that theories command our attention to the extent that they gain empirical support. A well-formed theory that has not gained empirical support can generate derivations. However, theory without empirical support cannot offer those derivations as explanations or plausible predictions. Are a theory's predictions and explanations limited to the range of phenomena over which it has been tested? Not at all! Theories are never limited by previous tests. To the contrary, the purpose of theory is to reach beyond what is known. However, the weight given to a theory's broader explanations and predictions is related to its success at surviving tests.

Empirical support of theories must necessarily fall short of proving them true. For example, had research supported Hypothesis 1 (or Working Hypothesis 1 from Chapter 1), that research would not prove it true. To do so, we would have to study *all* centralized structures that existed in the past, that exist now, and that could ever exist. However, as we have shown, the hypothesis can be proved false, as can any theory. In fact, the few cases we describe above suggest that Working Hypothesis 1, Hypothesis 1, and the Theory of Positional Centrality are false.

Having stated that evidence suggests that the Theory of Positional Centrality is false, two important qualifications must be added. First, falsification cannot rest on only a few isolated observations. Falsification also demands replication. Happily, the ability to replicate is a great strength of theory-driven experimentation while the inability to replicate is a great weakness of other methods of investigation used in sociology. For example, the exchange and centrality experiments discussed earlier are easily replicated and extended to other structures—at least in the low-tech setting—by anyone reading this book. In fact, any reader can also replicate the high-tech version of the experiments by using software that is available on the Internet. Later in the book we will tell how to access that software.

A second point must be made about falsification. At times, theories are only counted false if better theories exist to replace them. Whether better theories are needed, or whether only falsifying evidence is needed is always a question of what will be lost if the theory is set aside. The Theory of Positional Centrality could be falsified by evidence alone because it had (1) no prior empirical support and (2) it had no empirical implications beyond the one being tested. By contrast, any theory with substantial prior empirical support and extensive implications beyond the ones being tested is a theory with considerable utility. Instead of giving up all of that theory's ability to predict and explain after finding some falsifying evidence, new scope conditions are drawn that include the phenomena where the theory is effective and exclude the phenomena where it is not. Then only after a new theory with broader scope appears is our current theory considered falsified.

Theories As We Encounter Them in Sociology

The experimenter's task would be straightforward if all theories found in sociology were formalized so that testable hypotheses fell into our hands. Some theories are written in formal language and employ the language of logic or mathematics for precision. For those theories the experimenter's task in drawing hypotheses is straightforward. As we show in later chapters, such theories are also very useful in developing the experimental designs needed to test them.

In sociology, theories take many other forms. The most useful of these are, like the classical theories of Marx and Weber, grounded in extensive investigations of historical and contemporary societies. Such theories can be a valuable source, not just for hypotheses, but for conceptions of society on which formalisms are founded and experiments built. Formal theories that capture something of that classical understanding of society are covered in later chapters. Some social science theories mix assertions about the world and political ideology—certainly Marx did. In such cases, the researcher is challenged to separate the part of theory that can be tested from the ideological part that cannot. Other work focuses on what is called "metatheory." Metatheory includes presuppositions that underlie theories. Because presuppositions are not testable, metatheory need not concern us here.

Theoretical Research Programs: From Answers to New Questions

Many social scientists create or contribute to research programs. That is, they plan and carry out a series of related studies with each study adding to knowledge gained previously. Two types of research programs have been distinguished. A *cumulative research program* is a series of interrelated studies where each study in the series improves researchers' capacities "to identify and solve [scientific] problems" (Cohen 1989:293). Cumulative research programs like the one begun by Bales (1950) add to the storehouse of descriptions, classifications, or empirical explanations. However, the best-developed and most advanced sciences are those that generate and sustain *theoretical research programs* (Wagner and Berger 1985). A theoretical research program is a theory or set of interrelated theories and the research that tests them. These programs may also include studies that apply theoretical knowledge to concrete situations and problems (Wagner and Berger 1985:705).

Successful cumulative research programs can lead to theoretical research programs. Researchers can use a sequence of Type I, II, and III questions and the answers they find to establish first cumulative and then theoretical research programs. We use Bales' research to illustrate the process. As described in Chapter 1, Bales used his early observations to develop a system for describing and classifying interaction. His research asked and answered a series of Type I questions. What is an "act?" What kind of act did Subject 1 direct toward Subject 2 at 1 minute and 30 seconds into the session? Is that act similar to or different from the act that Subject 4 directed to the entire group at 4 minutes and 11 seconds into the session? While Bales' system for recording and coding interaction was an important development, it is most significant because it permitted him to observe behavior systematically and to ask more complex, Type II questions.

As already mentioned, Bales' system for describing interaction permitted him to observe that time spent talking is not divided evenly among members of task groups. We can imagine the research team asking the following question: "What things (or factors) are correlated with the time group members spend talking?" It is a Type II question. Today, we know that group members' perceptions of who contributed the best ideas, who did the most to keep the group working at the task, and who they liked best are all positively correlated with time spent talking. Yet they could not have found these answers had earlier studies not answered the Type I questions we described. Far too many programs end at this level of development (i.e., as cumulative research programs), but Bales pressed on to create a theoretical research program and many of his former research assistants went on to establish separate—but related—theoretical research programs. One such program was established and nurtured by Joseph Berger and his colleagues.

Birth of a Theoretical Research Program

Joseph Berger worked in Bales' laboratory in the early 1950s and, with several fellow graduate-student associates, went on to develop Expectation States Theory and the related Theory of Status Characteristics and Expectation States. We now describe a pattern of behavior uncovered in Bales' research and the Type III question that led Berger to develop Status Characteristics Theory (Berger, Fisek, Norman, and Zelditch 1977).

A typical Bales group developed a clearly identifiable interaction structure in about 20 minutes. However, Bales' assistants noticed an important anomaly. Some groups developed clearly defined interaction structures almost instantly. It was as if they had a group structure before they entered the laboratory. How could this occur?

Bales was very careful to assign complete strangers to experimental groups. Lack of familiarity with other subjects was an initial condition we listed in our description of the studies. After systematically reexamining their research protocols and their data, Bales' research teams discovered that the complete strangers thrown together in their laboratory waiting room did what complete strangers often do when they come together in a strange place. They talked to each other.

Strangers-in-waiting, having started conversations with those sitting near them, soon learned that "Tom Jones" was a sophomore, undecided about his major, and living in Lowell House. It was the last piece of information that turned out to be crucial. Upperclassmen at Harvard College lived in one of 12 residential houses—commonly called dormitories or residence halls at other colleges and universities. The houses were ranked according to social status and the status rankings were common knowledge among undergraduates. With some houses having higher status and others lower, the subjects developed status structures that reflected the rankings of their places of residence. Students who lived in the most prestigious houses talked more than students who lived in less prestigious houses. This discovery by Bales and his research associates uncovered an answer to the Type II question: What factor(s) is related to the time group members spend talking? One answer is residence-hall status.

The story has an important moral. Experimenters must be on guard because conditions they fail to control can inadvertently (and without their knowledge) produce the phenomena they are studying. Bales avoided recruiting subjects from the same house because he wanted to begin each experiment with students who were strangers. Experimental control ensured that subjects in each experimental group came from *different* houses. However, once subjects exchanged information about their residences, differences in residence-hall status affected group interactions and produced the rapidly developing structures he observed. Did other groups develop the same kinds of structures because they were also composed of subjects from stratified systems of residences? Did their group structures develop more slowly because subjects only came to learn about their differences in residence-house status (or other status differences) later in the experiment? Whether the evidence exists from Bales' studies to answer these questions we do not know, but we do know that a replication today could answer them.

Joseph Berger recognized that the phenomenon was more general than residence halls and undergraduates at Harvard. People from higher status

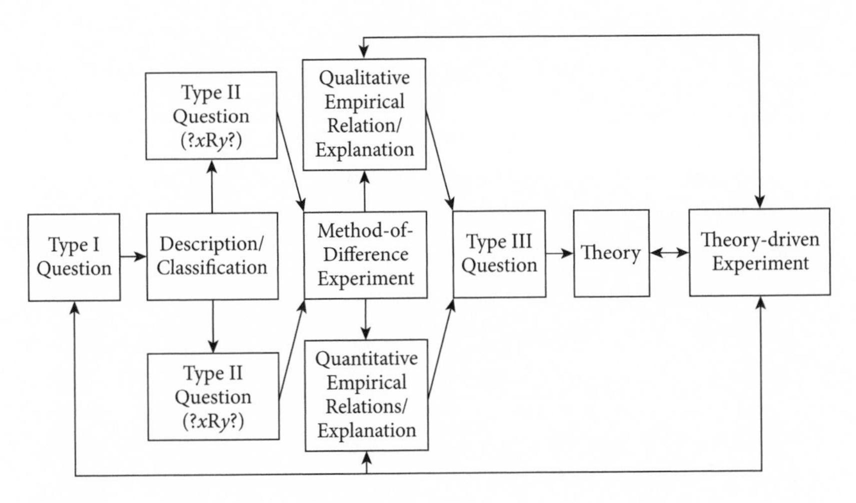

FIGURE 2.1. Questions, Answers, and Experiments in the Research Process

groups usually hold high rank in new groups containing persons from many different status backgrounds. A classic example is the finding that jury foremen typically hold higher status jobs than other jury members (Strodtbeck, James, and Hawkins 1957). Berger raised the Type III question (see Berger, Cohen and Zelditch 1966, 1972; Berger, Fisek, Norman, and Zelditch 1977):

> What accounts for the positive relation between the social statuses people hold outside a group and the status, power, or prestige of positions they occupy within the group?

Berger and his colleagues developed Status Characteristics Theory—a theory we will discuss in Chapter 4—as an answer for that question. We mention it here to show how the Type III question that Berger et al. (1977) raised can be tied to the Type I and Type II questions that Bales and his associates asked after observing elementary group processes over several years. We display the connections between types of questions and other elements of the research process in Figure 2.1.

Berger, Cohen, and Zelditch (1966, 1972; Berger, Fisek, Norman, and Zelditch 1977) built a theoretical research program to explain a phenomenon first isolated and identified in the cumulative research program begun at Harvard. This new theory had to be tested, and successful tests, by uncovering new phenomena unexplained by the original theory, raised new questions. To answer those questions, new and more complex theories were created and, when tested, discovered further phenomena thus raising questions yet again. The interrelated theories and the experiments that tested them are the nuts and bolts of a theoretical research program that remains vibrant today—almost 50 years after Berger completed the first studies as his doctoral dissertation (Berger, Fisek, and Norman 1998; Berger and Fisek 2006).

Readers acquainted with the history of the sciences know that science builds on previous work. It must, for if it did not, the same theories and explanations would be discovered and rediscovered time and time again. Still, building a science can take many forms. For example, not all theoretical research programs are, like Status Characteristics Theory, grounded in previous experimental work. Elementary Theory (Willer and Anderson 1981), a theory of social structure and power, builds on the historically grounded theories of Marx and Weber. Parts of Elementary Theory agree with parts of Marx and Weber.

Building on prior work, however, does not require agreement with it. As seen repeatedly in the physical sciences, some striking advances develop out of the sharpest disagreements. In Chapter 4, we review Galileo's study of falling bodies to show parallels between it and experiments in sociology. Importantly, Galileo's investigation revolutionized physics because the point of departure for the theory that guided his study was in sharp disagreement with Aristotle's theory of falling bodies. Yet Galileo as assuredly built on Aristotle by disagreeing as Newton built on Galileo by agreeing with him.

By now we have examined the kinds of questions that trigger sociological research and traced how Type I and II questions can build to Type III questions that motivate theory development. We have described what theory is and how the abstractions of theory are connected to the observable world so that they can be tested. Experiments offer the best tests of theory but not all experiments are designed to test theories. In fact, none of the three experiments described in Chapter 1 was designed to test theory. They are all empirically driven experiments and it is to an in-depth discussion of such experiments that we turn.

3 EMPIRICALLY DRIVEN EXPERIMENTS

THIS CHAPTER IS THE FIRST OF TWO THAT EXPLORE THE LOGIC of experimental designs. Here empirically driven experiments are analyzed. This type of experiment does not test theory. Instead, it is motivated by Type I and II questions and has as its objectives the discovery of new phenomena and finding relations between phenomena. Indeed, we will show that empirically driven experiments are powerful tools for discovering phenomena.

Our examination of empirically driven experiments begins with J. S. Mill's ([1843] 1967) canons of proof—the logical foundation of empiricist experimentation.[1] We introduce Mill's canons but focus on one—the method of difference—as the exemplar of the approach. We identify problems with Mill's conception of the scientific method and discuss R. A. Fisher's (1935, 1956) use of inferential statistics to revise it. We explain why Fisher's refinement of Mill's method was necessary, and identify criteria for the design of good empiricist experiments. Finally, we critically analyze the designs of four example experiments. One goal of our analysis is to demonstrate the power and range of empirically driven experiments.[2] A second goal is to identify the characteristics of good designs and to show how good and bad designs differ. The third goal is to explain how poorly designed or poorly conducted experiments can be better designed and more effectively conducted.

Mill's Canons and the Method of Difference

The work in which Mill proposed five canons for discovering regularities in nature is titled *A System of Logic* ([1843] 1967)—and for good reason. It systematically presents Mill's logic of experimentation and describes his methods of discovery—the methods of agreement and difference.

Mill's *method of agreement* is used to establish what logicians call "necessity." Assume that *A*, *B*, *C*, *D*, and *E* are potential causes and that *X* occurs in each of the following situations: (1) {*A*, *B*, *C*, *D*, *E*}, (2) {*A*, *C*, *D*, *E*, *F*}, (3) {*A*, *F*, *G*, *H*, *I*}, and (4) {*A*, *B*, *G*, *H*, *I*}. *A* is the only potential cause on which all four situations agree. *B* is absent in situations 2 and 3, *C* is not present in situations 3 and 4, and so on. Therefore, *A* is necessary for *X* to occur. Necessity can be written in logical notation, $X \rightarrow A$, which is read, "if *X* then *A*," but empirically, "*X* only if *A*," or "When *X* occurs, *A* precedes it."[3]

Mill's second method, the *method of difference*, is *the* method of experimentation. Mill identified it as a "more potent instrument" of discovery than the method of agreement. He stated the canon—the rule—as follows:

> If an instance in which the phenomenon under investigation occurs, and an instance in which it does not occur, have every circumstance in common save one, that one occurring only in the former; the circumstance in which alone the two instances differ is the effect, or cause, or an indispensable part of the cause, of the phenomenon. (Mill [1843] 1967:256 italics removed)

The method of difference is intended to establish "sufficiency." Consider an experiment that establishes two situations, the first of which includes the conditions {*A*, *B*, *C*, *D*, *E*}. The second consists of the conditions {*B*, *C*, *D*, *E*}. *X* is observed after the first conditions are observed but not after the second. *X* follows an occurrence of *A* but is absent when *A* is absent. The observations justify the statement $A \rightarrow X$ which is read, "If *A* then *X*." Empirically, it could be read, "When *A* occurs, *X* will follow." *A* is a sufficient condition for *X*.

Experimental control permits researchers to determine the "causal ordering" of phenomena (i.e., to determine whether *X* is a cause or an effect). In our example, an investigator would infer that *A* is a cause of *X* since *A* is the only difference between the conditions and it precedes *X*. Today, *A* would usually be called the independent variable and *X* the dependent variable. Here and elsewhere, independent variables are assumed to occur before and to cause dependent variables. Method-of-difference experiments like our example typically have at least one *control group* to which one or more *experimental groups* are compared.

Mill identified three additional canons: the *combined method of agreement and difference*, the *method of concomitant variation*, and the *method of residues*. The combined method of agreement and difference is just that—the joint use of the two methods just discussed. Consider two or more situations that include *A* ({*A*, *B*, *C*} and {*A*, *D*, *E*}), that are followed by *X and* two or more situations without *A* ({*B*, *C*} and {*D*, *E*}) that are not followed by *X*. The combined method of agreement and difference implies that *A* is a cause of *X*.

The method of concomitant variation compares varying values of *A* with varying values of *X*. If values of *X* vary consistently as *A* is varied, then *A* is a cause of *X*. Finally, the method of residues is applied when multiple causes or multiple effects may be present. The residue is what is left after all other causes have been determined. As an example, suppose a researcher observes the effects, *X*, *Y*, and *Z* under the conditions *A*, *B*, and *C* but does not observe them when all three are absent. If *A* and *B* are known to cause only *X* and *Y* respectively, the method of residues implies that *Z* is an effect of *C*.

We reconstruct Mill's reasoning as it is crucial to his idea of science ([1843] 1967). For Mill, experimental research finds regularities by *sorting them out* of an orderly world that, on the surface, gives the appearance of one chaos followed by another (248). Sorting out regularities is an *analysis*—in the literal meaning of that word—for it *cuts* the world into ever smaller parts. Why can regularities be found? Regularities can be found because (1) empirical reality is composed of regular patterns of relationships between phenomena, but (2) outside the laboratory many of those regularities are mixed to form the chaotic reality we observe. There are regularities *underlying* reality, but, until the regularities are sorted out and identified, the world of phenomena appears irregular (206).

The sense of Mill's system is this: Experiments cut the general regularity of the observable world—a regularity that cannot be directly observed—into precise, smaller regularities that can be observed. For Mill, statements about these regularities are scientific laws. Therefore, experiments discover scientific laws.

For Mill, the $A \rightarrow X$ regularity is not established by once observing *X* following {*A*, *B*, *C*, *D*, *E*} and once observing that *X* did not follow {*B*, *C*, *D*, *E*}. *An empirical regularity is exactly that—a series of events that will occur again and again in exactly the same way under exactly the same conditions.* Therefore, evidence for empirical regularity is demonstrated through replication—through running exactly the same experiment again and again with exactly the same results. Importantly, it is in the inability to exactly replicate that Mill's methods fail.

Mill's method fails to find laws for three reasons. First, his canon of proof demands perfect regularity. To find that A causes X, X must always follow A and never follow when A is not observed. In fact, Mill's method has never found a perfect regularity. The issue here is not the inexactitude of social science research. *No research in any science has ever satisfied Mill's requirement of perfect regularity.* Given measurement error alone, it is clear that perfect regularities cannot be observed.

Second, Mill's difference criterion cannot be satisfied. That it cannot follows from the statement, "every circumstance but one," in his description of the method of difference. The requirement has no practical solution. What it means can be seen in the following example. Mill asks the reader to imagine two circumstances:

$$ABCDE \rightarrow x$$

and

$$BCDE \rightarrow \sim x$$

Read the first as "When *ABCDE* are present, x—a measure or indicator of X—is observed" and the second as "When *BCDE* are present x is not observed." Because only A is missing in the second instance, the two instances of his hypothetical example have "every circumstance in common but one."

The task for the experimenter, however, is to create two *empirical* circumstances in the *real world* that are known to be exactly alike but for one and only one difference. Let us say that an experimenter tries to do so and has run the first of the above conditions and then the second. But now there are two differences because the time in which the two were run is different. More generally, any two empirical instances differ from one another, not in only one way as in Mill's hypothetical example but in many, many ways, any one of which could be claimed as the cause or an indispensable part of the cause of the phenomenon. It follows that Mill's is not a method that can actually be practiced.

A third problem confronting Mill's method is the inability of researchers to use observations as laws. Observations are always specific to times and places, and statements that describe them use concrete terms. On the other hand, laws are general statements that contain only theoretical concepts (Willer and Webster 1970). They are statements about relations between general ideas, not about concrete relations between observable phenomena. No system of logic permits a researcher to induce (or generalize) lawful statements from one or a thousand observation statements. In fact, that is why theories that contain laws must be tested. A scientist must determine whether the relations described by theories can be reliably reproduced.

Readers should greet every part of Mill's system for finding laws with skepticism. The world appears chaotic without analysis, but what evidence is there that there are underlying regularities and that those regularities jointly produce that chaos? There is none. Whereas the logic of Mill's method of difference requires perfect regularity, those who followed Mill were horrified to learn that, at times, *X* fails to occur after *A*, *B*, *C*, *D*, *E* occurred—and that *X* sometimes occurs when only *B*, *C*, *D*, *E* have occurred. Because method-of-difference experiments produce results that are inconsistent with Mill's difference canon, those who adopted his ideas about science had to find a less stringent rule.

Fisher and the Likelihood of Difference

Ronald A. Fisher (1935, 1956) used probability theory and inferential statistics to rescue Mill's method-of-difference experiment from the impossibility of discovering perfect regularities. Actually, Mill had already suggested that his methods could discover imperfect or probable regularities of the form: *A* is a probable cause of *X*. But probable regularities were important only as "steps towards, universal truths" (Mill [1843] 1967:387). By contrast, for Fisher, finding probable regularities was the goal.

Fisher made important contributions to experimental design including random assignment to control and experimental groups. For Fisher's studies in agronomy, random assignment meant the random assignment of one or another strain of wheat to one or another plot of ground. In the social sciences, random assignment means the random assignment of people to experimental and control groups. In either case, random assignment forms the logical foundation for the application of statistical tests to experimental data. Statistical tests evaluate whether independent and dependent variables are linked, while random assignment rules out at least some alternative causes of the dependent variable.

Let us reconstruct Fisher's reasoning. Random assignment to experimental and control groups permits the inference that the means and variances for any given measure are similar for the two groups *at the point of assignment.* With random assignment, members of the population from which the assignments are drawn are equally likely to be assigned to either group. As a result, the means and variances for the two groups should be nearly equal to each other. (They should also be equal to the population mean and variance though both are typically unknown.) According to the law of large numbers, that similarity should increase as the group sizes increase.

Consider a pool (or population) of potential subjects that consists of the 22,400 full-time undergraduates at the University of Arizona in spring 2006. Of this number, 53 percent are female and 28 percent are members of minority groups. Now let a researcher randomly select 200 members of the pool and randomly assign them to either a control group or an experimental group. Each group of 100 should have about 53 (53 percent) women and 28 (28 percent) minorities—and similarly for other characteristics. Moreover, the similarity in percentages for the two groups should increase if the group sizes are increased from 100 to 200.

Experimental and control groups in method-of-difference experiments are typically distinguished by an intervention called a *treatment.* The treatment is administered to the experimental and not to the control group. For example, in the *American Soldier* studies conducted during World War II (Stouffer et al. 1949), soldiers were randomly assigned to two groups. The experimental group saw a film, "Why We Fight," while the control group did not. The film was concerned with the evils of fascism and a comparison of mean responses on questionnaire items showed that the film had the hoped-for effect. Soldiers in the experimental group expressed significantly greater willingness to fight Germans than those in the control group. Because soldiers were randomly assigned to experimental and control groups, it was inferred that there would have been no difference in soldiers' willingness to fight Germans if none had seen the film.

Fisher's redesign of difference experiments introduced an important shift in the logic of discovery. As described above, the practice of randomly assigning subjects to experimental and control groups is tied to the assumption of equality of sample means. The *key* hypothesis under test, called the *null hypothesis*, is that there is *no difference* in sample means. More precisely, the null hypothesis asserts that, in the absence of treatment effects, if differences between sample groups are observed, they are chance differences produced by random assignment.

Fisher redesigned Mill's difference experiment. First, he abolished Mill's requirement that experiments find absolute regularity. Next, he replaced Mill's canon with the weaker rule of likely or probable regularity. In order to put the weaker rule into practice, Fisher had to overcome a problem. How is the likelihood of regularity to be determined? Probability theory seemed to offer a solution.

From the point of view of probability theory, subjects assigned to experimental and control groups in any experiment can be thought of as two assignments drawn from a (hypothetical) universe of all possible assignments. For that universe there is a (hypothetical) mean, μ, and variance, σ. If researchers knew that mean and variance, they could devise a statistical test to determine the likelihood that mean differences between experimental and control groups are due to the treatment. But the variance is not known nor can it ever be known for it is the variance of a hypothetical, not a real, universe of assignments. Probability theory's attempt to solve the problem seemed to lead to a dead end.

William S. Gossett, an enterprising statistician who worked for the Guinness brewery, devised a solution that substitutes sample variances for the unknown—and unknowable—population variance. Gossett's technique uses sample data to calculate standard deviations for each group, a calculation that may be familiar to many of you. An estimate of the standard deviation of the universe of assignments, the *standard error*, is calculated by weighting standard deviations by sample size. Then the standard error is used to calculate a t statistic. The value of t is the mean for the experimental group minus the mean of the control group divided by the weighted average of their standard deviations.[4] For samples of given sizes, Gossett's t statistic is used to determine the likelihood that mean differences could have occurred by chance. Recall, that the means of the experimental and control groups are presumed to be equal (or nearly so), before the experimental treatment. Thus, values of t that are unlikely to have occurred by chance (e.g., fewer than five times in a hundred) are used to infer that the experimental treatment is a probable cause of the effect.[5]

Fisher revised Mill's method of difference by abandoning Mill's criterion of absolute regularity and replacing it with a new criterion: *An observed difference is evidence of a probable regularity if it is unlikely to be due to chance.*

The purely arbitrary values of .05 and .01 have become the standards for unlikelihood in the social sciences (Leahey 2005). As an example, for the American Soldier experiment, Fisher would only require that significantly more of the experimental than of the control group wanted to fight Germans. By contrast, Mill would demand that all of the experimental group would want to fight Germans and that none of the control group would be so motivated. But results like those demanded by Mill are never found. Thus, Fisher's method, unlike Mill's, is workable. In fact, when *n* is large enough, sample means can be very close to each other and results may overlap substantially. Yet, a *t*-test may still imply that the independent variable is linked to its dependent variable.

Though Fisher's criterion has been substituted for Mill's canon, the objectives of empirically driven experimentation remain remarkably unchanged. The main objective is to produce science by applying the method of difference to find regularities. Now, however, the regularities are not absolute and certain, as demanded by Mill, but partial and unlikely to be due to chance.

Still, between Mill and Fisher some objectives have changed. Somewhere between Mill and Fisher, the goal of discovering laws was lost along with the hope of reconstructing reality out of discovered regularities. That they were lost suggests that researchers cannot use method-of-difference experiments to build science from observations. What we will now suggest is that building science is not the right goal for the empirically driven method-of-difference experiment.

Criteria for Empirically Driven Experiments

The empirically driven experiment, as refined by Fisher, has proved to be an important scientific tool because it is a remarkably effective method for hunting down relative regularities. Of course, its effectiveness for any particular study depends on whether it is used properly. In this section we identify criteria that empirically driven experiments must meet to be effective. If these criteria are not met, the results and their interpretation can be affected by *artifacts*. Artifacts are systematic effects produced by (1) features of the research setting, (2) the experimenter, or (3) subjects in an experiment. Artifacts can lead researchers to make errors in their inferences about the causes of observed effects (Kruglanski 1975:103). They may incorrectly attribute an experiment's effects to the variable under experimental control or fail to attribute effects to variables not under control—or both. The issue here is *internal validity*, whether the experiment has eliminated alternative causes of the phenomenon so that effects can be properly attributed to the variable under experimental control.[6]

The criteria for properly designing method-of-difference experiments follow directly from Mill's and Fisher's logic of experimentation. We put them in the form of four maxims:

1. Create at least two study conditions that are initially as identical as possible.
2. Introduce a single difference between the two conditions and observe the result.
3. Restrict all inferences about the result to the effect(s) of that single difference.
4. Infer relative regularities only if it is unlikely that they are due to chance.

In experimentation involving people, the first two criteria can be satisfied in two very different ways. First, the most frequently used procedure is to assign people at random to experimental and control groups so that subjects are initially identical except for differences that occur randomly in the assignment process. The American Soldier experiment took this form. Next, introduce the difference of interest as an experimental treatment. The film, "Why We Fight," was the crucial difference in the American Soldier study. Subjects in the experimental group saw it while soldiers in the control group did not. More elaborate designs, as in the Asch (1958) experiments discussed later in this chapter, have multiple experimental groups.

There is a second way that maxims 1 and 2 can be satisfied. Begin with two groups known to be different and study them under identical conditions. In this way, the researcher introduces the difference by studying two (or more) different kinds of people. For example, in Simpson's (2003) experiment discussed below, all subjects played exactly the same experimental games. The difference was that one group was male and the other female.

Satisfying the third criterion should be straightforward, but many experiments fail to satisfy it. For example, in a series of experiments discussed below, Milgram (1965, 1974) claimed that the experimenter's legitimate demands explain the willingness of subjects to administer severe electrical shocks to confederates. But Milgram's experiments did not systematically vary legitimacy and he had no control group in which legitimacy was absent. Finally, satisfying the fourth criterion, that relative regularities are unlikely to be due to chance, like the first three, can only be satisfied by well-designed studies. Now we analyze the designs of four method-of-difference experiments applying the four maxims.

The Asch Conformity Studies

Solomon Asch's studies of conformity and distortion of judgments are widely considered to be classic experiments. Asch was motivated by a Type II question. His objective was to discover social conditions (factors) that would "induce individuals to resist or to yield to group pressures when the latter are perceived to be *contrary to fact*" (Asch 1958:174). Writing in the aftermath of fascism and World War II, Asch asserted that understanding why people submit to group pressure is of "obvious consequence for society" (174). His experiments are still studied today. We will evaluate key elements of his designs against the maxims for good designs.

Asch's earliest study had eight men seated in a row (or semicircle) but seven of the eight were confederates of the experimenter. The men were seated facing stimulus materials used in the experiment. The stimulus materials were 18 pairs of cards that research assistants held up in sequence.[7] The first card of each pair contained a single vertical line called the *standard line*. The second card contained three vertical comparison lines. One comparison line was always identical to the standard line. The other two lines were clearly shorter or longer than the standard. The men's task was to judge which of the comparison lines was identical to the standard line and to report their judgments to the research staff.

Asch's design is a classic difference experiment. In the experimental treatment, the men were required to report their judgments aloud and in sequence. The lone subject was seated in either seventh or eighth position. For 12 of the 18 stimulus pairs, every confederate gave the same erroneous answer. Then the subject was asked to give his answer, also aloud. The "errors" given by the unanimous majority were intended to create social pressure, the independent variable. Asch used the subject's response, whether correct or an error, as the dependent measure.

Asch's control condition had the subject and all seven confederates write out their judgments (i.e., their judgments were kept private). Otherwise, the experimental and control conditions were similar. The design permitted Asch to measure the control group's error rate and compare it to that for the experimental group. The difference measured the effect of social pressure.

In the face of a unanimous majority, the 50 subjects compiled a 32 percent error rate and all errors were in the direction of the majority's responses. In the control group 37 subjects had an error rate of less than 1 percent (approximately .68 percent), a figure more than 31 percentage points lower than the experimental group. Subjects in the experimental group averaged 3.84 errors while those in the control group averaged only .08 errors each!

Asch studied conformity that occurs when people change their judgments to agree with those of others that are contrary to fact. It is important to understand the effect produced in Asch's experiment and how the research design allows a precise interpretation of that effect. Asch's concern was not whether subjects were capable of making accurate judgments. In fact, line lengths were selected to be obviously different and thus to produce the negligible error reported for the control group. Instead, the concern was what subjects would *say* their judgments were when they faced a unanimous majority who reported judgments that were obviously contrary to fact.

The design is excellent in that it clearly isolates power from influence—two phenomena that often occur together—and are often confounded. Though Asch used the term "influence" in the titles of some papers, influence occurs only when there is a change of beliefs (cf. Zelditch 1992). Asch's post-session interviews with subjects showed that they were not influenced: they rarely if ever believed that their erroneous responses were correct.

Asch was as interested in resistance to group pressure as in conformity to it. To investigate resistance, an array of contrasting experimental group designs were run with different independent variables. He varied the number of confederates, using 1, 2, 3, 4, 8, or 16. Whereas one confederate had almost no effect, the strongest effects were found for three and four confederates while effects leveled out or declined slightly with larger numbers. Applying Mill's canon of concomitant variation, conformity varied with the number of others who disagreed, a relation that was not linear, but curvilinear.

Asch also studied non-unanimous majorities and he did that two ways. Either a second experimental subject was substituted for a confederate or one confederate was instructed to make correct judgments. The error rate that had been over 30 percent for unanimous majorities, declined to 10.4 percent and 5.5 percent, respectively, for the two non-unanimous designs. Finally, Asch reversed the initial design by placing a single erring confederate in a group of 16 subjects. Far from affecting subjects' judgments, the confederate was the object of ridicule.

Asch's experimental designs have several good qualities. Independent and dependent variables were developed such that the independent variable (group pressure) could be varied and resulting variations of the dependent variable (the error rate) could be measured easily. Physical elements of Asch's study satisfy maxims 1 and 2 of the method-of-difference logic because the control and experimental conditions differ on only one factor—whether judgments were reported privately or publicly. Applying maxim 3, this feature strengthens the evidence for the link between the judgments given by confederates and the subjects' report of judgments that they knew to be wrong. However, Asch does not mention random assignment and only differences were reported. He did not use test statistics estimating the likelihood that differences were due to chance.

Two other good design qualities should be mentioned. First, designs like Asch's produce compellingly clean results by what we call the "sledgehammer approach." In the sledgehammer approach, the effect of the independent variable is so great that it overwhelms variation in the dependent variable due to sampling or measurement error. Asch's design produced strong effects and, as a result, left little doubt about whether group pressure affects judgments—in spite of the absence of random assignment. Second, the design was high in *experimental realism* in that subjects took the conditions of the experiment as both real and significant. Experimental realism is sometimes contrasted to "mundane realism," which means that the experiment is similar to particular conditions outside the lab.

The Asch experiments are still important today, at least in part, because many elements of the design were sound and because they produced very strong effects. But how good are his designs by contemporary standards and conventions? Today, random assignment would be required and quantitative data would be much more precisely presented. For example, he reports that the error rate in the control group was 2 percent of that reported for his first experimental group. The difference is very large and the effect appears obvious; but current standards demand, not that comparison, but statistical tests showing that the observed differences between various experimental groups were unlikely to be due to chance. The need for statistical testing comes to the fore in Asch's comparisons of groups of size 3, 4, and 5 where the differences between experimental groups are not so obvious. Journal editors might also demand further experiments to nail down the exact shape of the curvilinear relation between group size and the extent of conformity—something we would also like to know. Fortunately, Asch's sledgehammer-like effects and carefully crafted differences between experimental and control groups leave no doubt, at least for us, that the effects he measured were conformity created by group pressure.

The Milgram Obedience Experiments

We examine the design of Stanley Milgram's (1974) experiments on obedience to authority in this section. Descriptions of Milgram's studies are standard fare in college psychology textbooks; some high school psychology courses also discuss them. Like Asch, Milgram was motivated by a Type II question: What conditions affect obedience to authority (Milgram 1965)? Also like Asch, Milgram's work was motivated by issues raised during the tyrannical rule of Hitler's Third Reich. Many around the world asked how millions could be humiliated, tortured, and gassed in Nazi death camps. Was the Holocaust the work of many evil people or the result of the obedience of many ordinary people to the authority of an evil few? Milgram took the question seriously and designed experiments to discover the limits of authority.

Unlike many who run all their experiments on college campuses, Milgram conducted some of his studies at off-campus locations. He also departed from procedures used by most academic researchers by recruiting subjects from the general population, including factory workers, laborers, businessmen, professionals, and others. Students were specifically excluded (Milgram 1974:15). Furthermore, the Milgram studies, as we know them, were intended to be the first of a series of cross-national studies to learn how authoritarianism—and responses to it—varied. His U.S. studies were to be followed by studies in Germany where Milgram expected obedience to be higher. Milgram never replicated his studies in Germany, but Mantell (1971) found obedience rates as high as 85 percent in his studies of obedience in Germany.

The best known of Milgram's experiments, his Experiment 2, was designed as follows (1974:13ff). Two paid volunteers enter the laboratory. Only one is a subject while the other, unknown to the subject, is a confederate. The experimenter explains to them that the study of "memory and learning" involves two roles, "teacher" and "learner." The task required the teacher to read a series of word pairs to the learner. The learner was asked to reproduce the second member of each pair after seeing the first word. The teacher was instructed to give an electric shock to the learner after an error and to increase the intensity of shocks with each subsequent error. After the task was explained, a purportedly random drawing placed the confederate in the learner role and the subject in the role of teacher.

After watching the experimenter strap the learner into a chair or, as in some experiments, helping to strap the learner in, the subject accompanied the experimenter to a separate room. There the subject was seated at a "Shock Generator," the control panel of which had 30 switches controlling 30 levels of shocks numbered in 15-volt increments from 15 to 450 volts. Numbers on the Shock Generator were augmented with labels from "Slight Shock" to "Danger: Severe Shock." The last two levels (435 and 450 volts) were labeled "XXX."

No one is ever shocked in any of Milgram's experiments, but, beginning at 75 volts (shock level 5), the confederate (the learner) was instructed to follow a set pattern of utterances expressing increasing discomfort as the voltage is increased. At 150 volts, the confederate demands to be set free (Milgram 1974:56), at 300 volts he screams, and, as the Shock Generator approaches maximum voltage, he begs the subject to stop and refuses to answer further questions (Milgram 1974:57). The dependent variable is the maximum level of (notational) shock the subject administers before refusing to continue.

The experimenter took an important and active role in interaction with subjects. When a subject expressed doubts about continuing to higher voltages, the experimenter firmly asserted the following "prods":

Prod 1: "Please continue," or "Please go on."

Prod 2: "The experiment requires that you continue."

Prod 3: "It is absolutely essential that you continue."

Prod 4: "You have no choice, you must go on." (1974:21)

If the subject asked about harm to the learner, the experimenter said, "Although the shocks may be painful, there is no permanent tissue damage, so please go on" (21). That assertion was followed by prods 1–4 as needed to induce the subject to increase the shocks. If the subject explained that the learner did not want to go on, the experimenter was to say, "Whether the learner likes it or not, you must go on until he has learned all word pairs correctly. So please go on" (22). Then prods 1–4 were added as needed to induce the subject to increase the voltage.

Milgram ran 18 variants of this basic experiment and identified one (Experiment 11) as a unique "control" (Milgram's language). Subjects in the control experiment were allowed to select the level of shock. Milgram did not report whether prods were used in this experiment but prods might have had little relevance since the subject selected the shock level.

Subjects in the control condition gave an average maximum shock of 82.5 volts. Twenty-three subjects (57.5 percent) never achieved the 75-volt level at which confederates were to begin verbal reactions (72). Twelve (30 percent) set shock levels higher than 75 volts but only 1 subject (of 40) achieved the maximum shock level. By contrast, subjects in Experiment 2 described above administered an average shock level of 367.9 volts and 25 of 40 subjects (62.5 percent) gave the maximum shock of 450 volts. In fact, values for the control group are well below those attained in any other design.

There is no doubt that the Milgram experiments produced obedience but his is a poor design. Subjects were not randomly assigned to experiments, the independent variable is not properly measured nor varied, more than one variable varies across experiments, and Milgram never reports statistical tests. In fact, none of the maxims described above are satisfied in the series of experiments. For example, there are at least two differences between his control and Experiment 2. In the control group, subjects had control of shocks. By contrast, in Experiment 2, the experimenter specified shock levels *and*, if the subject hesitated, demanded that the subject give them.

Across all 19 experiments there is little or nothing in the designs that links obedience to authority. Authority is typically defined as *legitimate power* but Milgram offers the reader no definition of authority and the term "legitimate authority" was not used during the experiment. Certainly, the experimenter wore a lab coat and acted in an authoritative manner but what subjects inferred from the experimenter's demeanor or his position as a Yale professor (if known) is not known or reported. There were post-experimental interviews, but because no precise meaning was attached to legitimate authority, it is not clear how Milgram determined whether subjects responded to the experimenter's legitimacy or his coercive power.

Moreover, because authority was the independent variable, these experiments clearly violate maxim 2 above. As mentioned above, Experiment 11 was called a control but the variable controlled was not the experimenter's status as a legitimate authority. Milgram built the wrong control condition. Because his aim was to show that people acted out of obedience to authority, the control group should have been one in which a nonlegitimate, or an illegitimate experimenter made demands. But that condition was never run.

Milgram asserts that the experimenter was a legitimate authority figure but he mentions anecdotal evidence suggesting an absence of experimenter legitimacy. Experiments not run at Yale University were carried out in nearby Bridgeport, Connecticut, where the experimenter claimed to represent the entirely fictitious "Research Associates of Bridgeport." Remarks of two subjects in those studies questioned legitimacy. The first wrote:

> Should I quit this damn test? Maybe he passed out? What dopes we were not to check up on this deal. How do we know that these guys are *legit*. No furniture, bare walls, no telephone. (Milgram 1974:69, emphasis added)

The second proclaimed:

> I questioned on my arrival my own judgment [about coming]. I had doubts as to the *legitimacy* of the operation and the consequences of participation. (Milgram 1974:69, emphasis added)

Importantly, Milgram found similar levels of obedience in Bridgeport and at Yale.

There were other problems with Milgram's design. Like Asch's studies, Milgram's experiments had sledgehammer-like effects on the dependent variable, a good quality. However, unlike Asch's research, the independent variable was not nailed down when the experiment was designed. Were the effects caused by legitimate authority? Or did Milgram's subjects respond to coercive power?

Once the experiment began, one thing was certain: The subject had every reason to believe that the experimenter was absolutely ruthless. To get his data the experimenter strapped someone down and had him shocked to dangerous levels, perhaps to death. Moreover, on hesitation, that ruthless person firmly asserted, "You have no choice, you must go on" (the fourth prod). Whereas the prod did not specify the consequence of disobedience, coercive threats need not be explicitly stated. The *Law of Anticipated Reactions* (Nagel 1975) asserts that people comply with coercive power by anticipating both what the powerful person wants them to do *and* what the powerful person *will do to them* if they fail to comply.

Is coercion a more plausible cause of obedience than legitimate authority? Or was Milgram right that legitimate authority produced obedience?[8] Or was there some third or fourth cause? Because they were poorly designed, those questions cannot be answered by Milgram's studies.

Automaticity of Social Behavior

This section analyzes the designs of three experiments Bargh, Chen, and Burrows (1996) used to study *automaticity* of social behavior. Social behavior is "automatic" when it is triggered by objects or events present in a situation and when it is not mediated by conscious perceptual or judgmental processes (1996: 231). That is, the connection between triggering object or event and behavior is direct *and* unrecognized. The objective of the experiments was to uncover automaticity—to discover whether a stimulus that subjects fail to recognize can affect their behavior. Discovering or developing greater specificity of phenomena are Type I activities and linking stimuli to behavior is a Type II issue. Though Bargh et al. extensively cite related research and comments by the philosopher William James, theory played no part in the design of these experiments or in the formation of the hypotheses that motivated them.

All three of Bargh et al.'s experiments shared the following design elements:

1. Subjects were randomly assigned to experimental and control groups.
2. Priming conditions—the independent variable—were buried in a larger task, the completion of which was purportedly the experiment's aim.
3. Measurement of the dependent variable was disguised and, in two of three cases, the time of measurement appeared to occur either between two parts of the experiment or after the experiment.

The first experiment investigated whether politeness and rudeness could be primed such that a subject's behavior evinced one or the other *and* without the subject's conscious recognition that politeness or rudeness had been produced. Subjects were told that the experiment was divided into two parts, both concerned with language ability. The experiment had three priming conditions: polite, neutral, and rude. Each subject was given a "Scrambled Sentence Test" consisting of five scrambled words (e.g., "he it hides finds instantly"), and a list of 30 words that could be taken, one at a time, to construct

30 sentences of four words each. Of the 30 words, 15 differed across the three conditions. Words for the rude priming included, *bold*, *rude*, *bother*, *brazen*, and so on. Words for the neutral priming included, *exercising*, *occasionally*, *practiced*, *clears*, and so on. Words for the polite priming included, *respect*, *honor*, *yield*, *polite*, and so on. The three sets of 15 words represented variation in the independent variable.

The dependent variable was measured in the following way. Subjects were told, on completion of the test, to go into a hallway and find the experimenter so that the second part of the experiment could begin. In fact, the experimenter was posed at a partly open door apparently explaining the experiment at length to a second "subject" who was not visible to the first subject. The second subject was a confederate and the experimenter's explanation would continue until interrupted by the first subject or until 10 minutes had expired. The dependent variable was the time that expired before the subject interrupted the experimenter.

The researchers identified effects in two ways. First, they compared mean times (in seconds) expired before interruption. Here are the results: rude = 326, neutral = 519, and polite = 558. Analysis of variance (ANOVA) showed a significant effect and the mean time expired in the rude treatment was significantly different from the other two as evaluated by the F-test. A second measure—the percentage of subjects who did not interrupt—was developed because 20 of the 35 subjects (57 percent) did not interrupt the experimenter and, instead, waited the full 10 minutes (i.e., 600 seconds). Fewer than 20 percent of subjects in the polite group interrupted the experimenter, about 35 percent of the neutral group interrupted and more than 60 percent of the rude group interrupted. These results are clearly consistent with the hypothesized priming effect. To assess whether it was an automatic effect or whether these behavior differences were mediated by interpretations of the experimenter's behavior, subjects were asked whether they found the experimenter to be polite. The experimenters report no group differences in subjects' evaluations of the experimenter's politeness.

The second experiment was designed to activate an "elderly" stereotype. Once again the Scrambled Sentence Test was used. In this case the experimental group was given words that might prime elderliness while the control group was given age-neutral words. Importantly, for both groups, no words concerning time or speed were included. The dependent variable was time taken to walk down the hall after the apparent completion of the experiment.

Measured in seconds, the experimental group averaged 8.28 while the control group averaged 7.30. The difference was significant beyond .01. An alternative explanation for the finding is that mood intervened. To test for mood, a new set of subjects took the same Scrambled Sentence Test followed by a test measuring emotional states. No significant differences in emotional states were found. Because time and speed were not mentioned to either experimental or control group and mood did not intervene, it was concluded that the effect measured was automatic.

The third experiment sought to measure effects of African-American compared to Euro-American stereotypes. No subjects were African American. Prior to a long and tedious task, a face was briefly shown on the subject's computer screen, so briefly that subjects could not recall seeing it. After subjects had been engaged in the task for some time, the computer asserted that all data up to that point had been lost and that the subject must begin again. A hidden camera watched by two coders, blind to the experimental conditions, rated the hostility of the subject's facial reaction. We mean by "blind to the experimental conditions" that the coders did not know the hypotheses being tested and did not know which face was shown. The set of subjects primed by the African-American face were significantly more hostile. Yet, subjects did not recall seeing the face. Racial attitudes of those in the experimental and control groups were tested. Because they found no differences between the experimental and control group on racial attitudes, the experimenters concluded that the effect was automatic.

These very well-designed experiments easily satisfy the four maxims listed above. Several design features contribute to that quality. First, random assignment was used to establish the likelihood that experimental and control groups are initially identical but for chance differences. Furthermore, random assignment satisfies assumptions for using statistical procedures to evaluate observed differences. Second, the experimental and control groups differed only on the priming conditions. However, it would be wrong for researchers to claim that any observed differences were automatic responses to priming events if subjects were aware of their purpose. Thus the researchers buried the priming events in a larger task. The instructions gave a cover story obscuring the experiments' goals. That procedure should have eliminated effects that might have occurred because of *demand characteristics* where the subjects infer the experimenters' hypotheses and try to ensure that they were satisfied.[9] Third, they concealed measurement of the dependent variable by measuring it at a time that did not appear to be part of the experiment. That procedure is undoubtedly necessary when a phenomenon as subtle as automaticity is the object of investigation. When measurement is effectively disguised, subjects are not able to help or hinder the experimenter by purposefully producing (or failing to produce) behavior they believe the experimenter expects or desires. Said somewhat differently, when subjects are unaware of the behavior that is measured, demand characteristics cannot affect that measurement. Fourth, the experimenters use of double-blind procedures reduces the possibility of experimenter bias—effects caused by experimenters whose behavior knowingly or unknowingly biases subjects behavior to support hypotheses.

Finally, the experimenters were not satisfied to produce effects and then hope that the reader would conclude from the experimental-control group design that the effects were automatic. To the contrary, interviews were used to identify and eliminate plausible factors that might establish a conscious rather than automatic link between the priming events and subsequent behavior.

Sex, Fear, and Greed in the Laboratory

The investigation examined in this section explores whether men's and women's motives differ, leading them to respond differently to certain social dilemmas. The idea is a Type II issue: Is the factor, *A* (gender), linked to *X* (response to a social dilemma)? Brent Simpson (2003) employed the method-of-difference logic in the following way. Conditions experienced by all subjects were identical: all played exactly the same three dilemma games. The only difference was whether the subjects were males or females. Thus the design satisfies maxims 1 and 2. Simpson's design differs from the three previous designs in a unique way: His is a hybrid design that combines qualities of method-of-difference (empirically driven) and of theory-driven experiments.

A brief review of some elements of game theory, as used in this experiment, will help show how theory influenced the design. Table 3.1 presents three games in "normal form." Each game is played by two people who decide simultaneously and without knowledge of the other's decision. For the experiment, each person is asked to seek a payoff as high as possible by selecting one of two options, labeled C and D. C stands for cooperation and D for defection. For each game, think of Person 1 selecting either C (the top row) or D (bottom row) and Person 2 selecting between the C (left column) or D (right column).

TABLE 3.1 Payoff matrices for three social dilemmas

	Prisoner's Dilemma		*Fear (No Greed)*		*Greed (No Fear)*	
	C	D	C	D	C	D
C	3, 3	1, 4	4, 4	1, 4	2, 2	1, 4
D	4, 1	2, 2	4, 1	3, 3	4, 1	1, 1

SOURCE: Simpson (2003).

All three games in Table 3.1 are *strategic decision situations*. By "strategic" we mean that the payoff to each person is jointly determined by the choices of both. Consider first the payoffs in the Prisoner's Dilemma (hereafter P/D) game. The payoff to the left of each comma goes to Player 1 (hereafter P1) and the payoff to the right of each comma goes to Player 2 (hereafter P2). For example, when both choose C (cooperate), both receive 3 points, but when P1 chooses C and P2 chooses D (defect), P1 receives only 1 while P2 receives 4. Reversing the choices reverses the payoffs: when P1 selects D and P2 selects C, they receive 4 and 1 respectively. Finally, when both defect, both receive 2. Examination of the other two games in Table 3.1 shows that both are also strategic decision situations because the payoff to each person is jointly determined by the choices of both.

What should a subject do to gain the best payoff from the P/D game? First ask, "What is the best choice when the *other* cooperates?" Then ask, "What is the best choice when the *other* defects?" Now compare the answers to the two questions. Are the answers the same? If so, that is the best strategy. Because the game is symmetrical, finding the answer for one player finds the answer for both.

Let us now walk through the procedure just explained from the point of view of P1. What should P1 do if P2 cooperates? When P2 cooperates, P1 will receive 3 by cooperating or 4 by defecting. Thus P1 should defect. What should P1 do if P2 defects? P1 will receive 1 by cooperating and 2 by defecting. Thus P1 should defect. The answers are both "defect." Therefore, P1 should defect regardless of the decision of P2. Because the game is symmetrical, P2 also should defect. More generally, when one option is always better regardless of the option selected by the other—as it is here—that option is called the *dominant strategy*.

It is now possible to see why the P/D game is called a dilemma. As just seen, it is rational for both players to defect and, when they do, each gains 2 points. The game is a dilemma because each would gain not 2, but 3 points were both to cooperate. The dilemma cannot be resolved when two people play each other only once—as they do in Simpson's experiment.[10]

Drawing from two informal perspectives, evolutionary psychology, and "role theory," Simpson hypothesizes that males and females defect for different reasons. A male playing the P/D game will defect because he is more competitive and more willing to take risks. Call this motive *greed*. A female playing the P/D game will defect because she is less competitive and less willing to take risks. Call this motive *fear*. The P/D game contains both fear and greed. Therefore, males and females should defect at similar rates.

Simpson built two additional games—Fear and Greed—each containing only the corresponding motive. Neither game has a dominant strategy. In the Greed game, defection is preferred only when the other cooperates. If the other defects, the player gains exactly the same payoff from cooperation and defection. Simpson hypothesizes that, in the Greed game, males should defect more than females. In the Fear game, defection is preferred only when the other defects. If the other cooperates, the player gains exactly the same payoff from cooperation and defection. Simpson hypothesizes that, in the Fear game, females will defect more frequently than males.

Now the combined use of the method of difference and theory in the experimental design can be seen. Game theory is used to create three social dilemmas. Are there gender differences across these dilemmas? Only a mixed design can answer that question for the following reason. Whereas game theory builds the dilemmas, it does not differentiate people by gender. In fact, game theory treats all humans as deciding in exactly the same way. Therefore, a difference design is needed to test hypotheses designed to discover the presence or absence of gender differences.

The experiments were carried out in two settings. For setting 1, subjects were seated at a PC and played against fictive partners who were purportedly seated in other experimental rooms. Subjects were told that they would play three games against three different partners. They were told that they would earn $1 for each point they gained in each game but they would not learn others' choices or their point totals until they completed all three games. In fact, there was no other player and subjects were paid a fixed amount for their participation. The second setting differed in two ways: (1) subjects gave their responses on paper, and (2) they were told to imagine that the points had value to them and others but money was not mentioned. The two settings were similar in every other respect as subjects in the second setting were also paid a fixed amount.

Simpson's study uses a 12-condition 3 (games) × 2 (settings) × 2 (gender) *factorial* design. Factorial designs (Campbell and Stanley 1966) allow researchers to make comparisons of multiple differences simultaneously. As an example, *statistical control* can be used to hold game and setting constant so that an analyst can compare female and male responses for the P/D game in the PC setting. Simpson was able to compare the effects of games (Fear, Greed, and P/D), settings (PC and non-PC), and gender (female, male). These comparisons are the *main* effects of the factors (independent variables). Factorial designs also permit evaluation of questions about *interaction* effects, the combined effects of two or more independent variables. In fact, Simpson's hypotheses involve interaction effects (e.g., whether females and males respond similarly or differently to the various games).

Results from both settings offer partial support for Simpson's hypotheses. As hypothesized, males and females defected at similar rates for the P/D games in both settings. For the Fear game, there were no female-male differences in defection for either setting. That finding ran contrary to Simpson's hypothesis that females would be more likely to defect. Finally, the results from both settings strongly supported Simpson's hypothesis for the Greed game. Males were substantially more likely to defect than females.

The series of experiments has many good design qualities that easily satisfy the four criteria for empirically driven experiments. First, although Simpson did not randomly assign subjects to experimental and control groups, he used a more powerful technique to ensure similarity of situations. *All* subjects completed all three games and, of the three, the P/D game acted as the control condition. Thus subjects in experimental and control groups were not just similar, as could be assumed under random assignment, but identical. Second, subjects played the identical games in two different settings and, third, within games in specific settings, the conditions were identical except for the subjects' gender. Game-specific differences in the responses of female and male subjects cannot be attributed to idiosyncratic differences in the sample. Because idiosyncratic differences are ruled out, we can be confident that observed differences reflect differences in the effects of games, settings, or gender. As an example, female and male differences in rates of defection for the Greed game are due to effects of the game *not* characteristics of the subjects.

Simpson's use of game theory to construct the P/D, Fear, and Greed games allowed the relative strengths of fear and greed motives to be calculated for each game (e.g., the difference is zero for P/D) and then precisely compared across games. Of equal importance, the three games define three distinct social dilemmas that mirror situations that occur outside the laboratory. For example, it is well established that problems of cooperation within and between societies very frequently take the form of P/D games (Axelrod 1984). There seems little doubt that future research will find societal problems of cooperation that take the forms of Fear and Greed games.

Furthermore, the findings are robust across two settings, thus reducing the possibility that the effects were artifacts of a particular setting. Unlike the Milgram experiments, the experimenter had no immediate role in producing the effects. And unlike the Asch or Milgram experiments, confederates were not used. Confederates, no matter how well trained, can affect subjects' behavior by introducing demand characteristics and experimenter bias through unintended variations in their behaviors (see Chapter 5). Neither the experimenter nor confederates introduced the contrasting experimental conditions. The games did. Because experimental conditions could not be affected by either subject-experimenter or subject-confederate interactions, few if any opportunities arose for introducing experimental artifacts.

Critical Review

The four experiments analyzed here show how difference experiments should and should not be built. The good designs discovered whether independent and dependent variables were linked by varying the absence, presence, or value of the independent variable and observing whether there were corresponding changes in the dependent variable. Experimental and control conditions differed in only one regard. Ideas about the phenomena under investigation determined which condition counted as control and which as experimental. In Asch's conformity studies, the control condition eliminated social pressure, whereas experimental conditions used differing degrees of social pressure. Simpson used the P/D game as the control condition for two other games. In fact, each of the other two games differed in only one way from the P/D game.

The four studies considered here could each be the point of departure for a series of additional studies. The Asch experiments show us that the size and unanimity of majorities affect conformity, but that is only a beginning. Many other factors may affect conformity including the status of confederates relative to the subject, the credibility of the majority's responses, and so on. Is a low status subject more likely to conform to the responses of seven high-status confederates than any of Asch's subjects where status was not differentiated? If so, how does relative status combine with size of the majority to affect conformity?

Milgram's experiments have been replicated in countries across the globe with similar results. They demonstrate conclusively that the capacity of humans to inflict pain on their fellows is substantial. Furthermore, Milgram found that factors like the physical and psychological closeness of the teacher and learner affect rates of obedience. Moreover, recent evidence that agents of the U.S. government and authorities in other countries have tortured civilians reemphasizes the research's importance. Yet, to our knowledge, alternative interpretations of the causes of obedience—including those we suggested—have not been ruled out. They have not been ruled out because no one has run a Milgram-type experiment in which the independent variable is known to be the legitimacy of the experimenter and only the legitimacy of the experimenter. The continuing notoriety of Milgram's designs coupled with the increasingly tight scrutiny of institutional review boards may make it more difficult to replicate them today.[11] Nevertheless, creative minds will find other ways to design new experiments in which control and experimental groups are properly designed. Then the effects that legitimate authority and coercion have on obedience can be distinguished and the effects of other motives ruled out.

Bargh et al.'s experiments were very well designed: they satisfy all four maxims for empirically driven experiments. As a result, they have made a strong case for the claim that behavior can be shaped automatically and without mediation by conscious perceptual or judgmental processes. Can automaticity have an array of other consequences not considered in their research? For example, even when fully rational, decision making can be affected by framing conditions established prior to decisions. Additionally, losses have greater disutility than the positive utility of gains of equal size. Can automaticity frame rational decision making? If so, we will have learned that preconscious processes unknowingly frame the most conscious of human activities, the making of decisions.

Turning to Simpson's study, follow-up work has already been done. Kuwabara (2005) suggests that negative results for the Fear game were the result of Simpson using the wrong game to test his hypothesis. A better test would devise a game in which fear of being exploited is a fear of the other's greed! Kuwabara built and ran that game: payoffs are a composite of fear of the other's greed. In effect, his experimental subjects played the Fear game against the simulated other's Greed game. Kuwabara found, as Simpson initially predicted, that females defected substantially more often than did males.

Looking back over the studies considered here, two experiments, Bargh et al. and Simpson were particularly well designed and offered exciting new discoveries. Despite some design shortcomings, Asch also offered exciting new discoveries. He found that the degree of unanimity was associated with the degree to which subjects conformed and offered evidence that there may be an optimal group size for producing conformity.

Milgram's experiments are, by far, the most poorly designed of the four. He produced sledgehammer-like effects that varied with the teacher's closeness to the victim and the experimenter. But was obedience affected by authority? Because authority was not separated from power (as Asch separated power and influence), it was never varied and nothing can be concluded about the effect of authority on obedience. Therefore, his claim to have discovered a link between legitimate authority and obedience is without foundation.

As we suggested at the outset of this chapter, well-designed empirically driven experiments have become powerful tools for discovering phenomena. That they have is well illustrated by three of the four experiments. The more than 50 years between the first and last studies clearly demonstrate a strong refinement of design, which promises to further discovery in the future.

The significance of these discoveries is strongly related to whether experimental findings can be applied outside the lab. When applying the results of empirically driven experiments, the more realistic, and the less artificial the experiment, the better. Realistic experiments are better because generalizing findings demands point-by-point similarities between the laboratory conditions and instances in the world outside the lab. A new example illustrates this point. Lucas, Graif, and Lovaglia (2006) hypothesized that, as severity of crime increased, prosecutorial misconduct (including withholding information from the defense) also increased. Lucas et al. created conditions much like those faced by prosecutors and their hypothesis was supported. Given experimental-prosecutorial similarities, generalization of their results is straightforward.[12]

By contrast, it is difficult to imagine encountering conditions outside the lab like those studied by Asch. Why did Asch design experiments with highly artificial conditions that block generalization outside the lab? He did so (1) to discover how group pressure affects levels of conformity when the subject had sure knowledge that his conforming responses were objectively wrong, (2) under such sparse conditions that the link between confederate pressure and subsequent conformity was clear. More generally, the simpler an experiment's conditions, the more certain a researcher can be of the link between independent and dependent variables. Yet, the simpler an experiment's conditions, the greater the difficulty in generalizing outside the lab. *For empirically driven experiments, internal validity and external validity are in opposition.*

Mill and those who followed him have long needed a more efficacious method than point-by-point similarity to apply results outside the lab. Such a method is implied by Mill's logic, but has never been developed. Following Mill, experiments are analytic in that they cut the world into simpler parts. Therefore, Mill should have devised a *method of synthesis* that recombines regularities into the complex whole that Mill postulated. Yet, neither Mill nor Fisher devised a method of synthesis and none has subsequently been put forward.

The next topic taken up is theory-driven experiments—investigations designed by theory. The logic of theory-driven experiments establishes a method for cutting the world into smaller, simpler parts that can be studied in the lab. It also includes a method of synthesis that resolves the problem of recombining experimental results to link experiments to naturally occurring phenomena. The results found in physics laboratories can be applied to galaxies many thousands of light years away. Do those same capabilities hold for sociology? We assert that they do and show, in the next chapter, that the methodology of theory-driven experiments in sociology and in physics is exactly the same.

4 THEORY-DRIVEN EXPERIMENTS

THIS CHAPTER EXTENDS OUR EXPLORATION OF THE LOGIC OF experimental design to theory-driven experiments. Theory-driven experiments test hypotheses derived from theory, but the role of theory is not limited to providing hypotheses. *Theories design theory-driven experiments.* Moreover, theory is the bridge that connects observations made in controlled laboratory environments to the world outside the lab. Because theory designs experiments, no single overarching design (e.g., control group vs. experimental group) exists for theory-driven experiments. Instead, experimental designs vary with the theories tested. Nevertheless, all theory-driven experiments share a common logic and, consequently, principles of good design.

This chapter is organized around a central idea: *Theory is the method of science—of all science.* In the first section, we describe the logic that underlies theory-driven research and identify principles of good design. To our knowledge, these principles have not been written down previously. We begin by analyzing two sets of experiments. The first are physics experiments that test Archimedes' laws of levers, the very first scientific laws ever proposed. The second experiments test simple social structural ideas. Through those experiments we uncover the principles of good design common across experiments in these two fields.

Without further proof, our claim that all sciences use the same method and share the same principles might ring hollow. We analyze four more experiments to show how pairs of theories with similar structures drawn from sociology and physics design similar experiments. The first pair is Galileo's investigation of falling bodies and James Moore's study of status-based influence. Each is a *founding experiment* because each was (1) designed by theory, (2) to test that theory, and (3) appears early in the systematic development of its science. The next two experiments are also founding studies designed by theories of geometric optics and network exchange.

Theoretic Science and Theory-Driven Experiments

Theoretic science understands the world in a way that is quite different from J. S. Mill, R. A. Fisher, and others who followed them. Theoretic science is not organized around the baseless assumption that the world is regular. Consequently, theory-driven experiments are not designed to uncover (presumed) underlying regularities. Nor is theoretic science built on the equally baseless assumption that the world is irregular. What, then, does theoretic science assert about the regularity of the world? It claims that whether the world is regular cannot be judged independently of the theories through which the world is understood. *Experience has shown that theory can make the world appear regular.* Does theory reveal underlying regularity? Or does theory only frame our perceptions? No one can answer these questions.

As pointed out in Chapter 3, because method-of-difference experiments are designed to discover regularities, they must be replicated. They must be replicated because a single observation is not a regularity. Although theory-driven experiments are not designed to find regularities, replication is equally important to theory-driven research. These considerations lead to two questions. First, what is replication in theoretic science? Second, because the purpose of theory-driven experimentation is not to find regularities but to test theory, why does science demand replication?

Scientists replicate an experiment when they test the same theory derivation several times and regularly produce similar results. Are results similar because the theory is powerful enough to produce regularity? Or does replication rest on underlying regularities uncovered by theory? Again, no one can answer these questions. Nor need they be answered because the importance of theories rests on their *confirmation status*, on the array of tests a theory has survived. In turn, their confirmation status rests only on the results of theory testing and not on unverifiable assumptions about the nature of the world.

Replication can have a second meaning: that an array of experiments has been carried out across a theory's scope of application. Here the experiments will not be similar, but all will be covered by derivations from the theory. Success in this kind of replication offers particularly strong support for a theory. As Popper pointed out in his discussion of the power of theories:

> Testability has degrees: a theory which asserts more, and thus takes greater risks, is better testable than a theory which asserts very little. (1994: 94)[1]

Replication of experiments has a central place in theoretic science for two reasons. First, if the theory is a good one, its experiments produce reproducible findings. Thus, replication is an important demonstration of the power of a theory. Second, replication is central because *scientific knowledge is public knowledge.* Those who devise a theory or conduct the first tests may have a personal stake in creating studies that find evidence to support their theories. However, because science is public, one researcher can check the work of another. Because others can scrutinize old results and produce new results, research will be more carefully and honestly conducted. As a rule, the science community may place greater confidence in theory supported by replications, at least some of which are carried out by disinterested third parties who have no stake in the theory's "success."[2]

Any discussion of replication in sociology and other social sciences may provoke a certain sense of skepticism because there are few if any replications. Replications are so few, in part, because most sociological research is descriptive or built on Mill's logic in a way that blocks replication. Surveys give information about a particular population of people at a particular place and time. After a survey is concluded, the time at which it was carried out is past and will never come again; therefore the survey cannot be replicated.[3] A case study of a particular group at a particular time and place also cannot be replicated. A critic of experiments might point out that no experiment can ever be run exactly the same way twice and thus claim that theory-driven experiments cannot be replicated. Replication of theory-driven research does not require a researcher to revisit the past but only to reapply the theory. Replication means only that the theory is applied again in the same way as before.

Because theory is the method of the sciences, theory-driven experiments must necessarily stand in the same logical relation to theory across the sciences. The unity of method offers a crucial advantage: One can learn how to design and conduct theory-driven experiments in sociology by studying theory-driven experiments in any other science. That is, social science experiments and experiments in other sciences are equally instructive. We examine six experiments below, three from physics and three from sociology. Understanding these experiments does not require prior knowledge of the physical theories being tested nor—for that matter—of the social theories. In each case, we describe the theory and then show how it was used to design experiments.

The experiments are presented in pairs, one each from physics and from sociology, and often there are greater similarities within each pair than across the pairs. We use these comparisons to show that (1) the logic that underlies theory-driven experimentation and (2) the forms that experiments take are the same across the sciences. The fact that theory-driven experiments share the same logic across the sciences has many important implications. We address those implications as the chapter unfolds but we mention two here. First, it is an often-heard criticism in sociology that experimental results cannot be generalized from the lab. The criticism is wrongheaded because theory-driven experiments do not generalize at all. As explained at the conclusion of the previous chapter, these experiments test theory. Once tested, it is the theory—not the experiment—that has implications outside the lab. In fact, results of sociological experiments apply outside the lab in the same way as do those of other sciences (Lederman 1993:101). How do we know that theories in other sciences apply outside labs? If physical and chemical theory applied only in the lab, your car would not start and your computer would not function. Indeed, neither you nor anyone else would have a car or a computer for they could not have been invented or manufactured.

The second implication of the unity of theory and theory-driven experiments across the sciences is that scientific progress goes on in sociology exactly as in other sciences. Science progresses by developing broader and more precise explanations and predictions based on broader and more precise theories. The most effective way that science progresses is through the interaction of theory and experiment. As examples found throughout the rest of this book will show, some areas of sociology are already progressive. (Also see Berger, Willer, and Zelditch 2005.) Like many other sciences, progressive research areas in sociology have advanced through the interaction of theory and experiment. Still, we must say again that experimentation is not a necessary condition for scientific progress—theory is. Since Darwin, and before the recent genetic revolution, biology has shown that a science can progress with little or no experimentation. Biology was progressive through the interaction of theory and the comparative method—but, arguably, that progress has been slower and more difficult than in theory-driven experimental sciences.

Laws, Models, and the Logic of Experimental Design

It is time now to uncover the criteria necessary to build good theory-driven experimental designs. We begin with the work of Archimedes. First, we design experiments to test Archimedes' laws of levers ([232 BC] 1897). Using those experiments as examples, we identify criteria for designing good theory-driven experiments. Then we design experiments to test the principle of rationality in social relations and apply the criteria to them. By doing so, we offer our first demonstration that the principles used to build theory-driven experiments are the same for the physical and social sciences.

Testing Archimedes' Law

Archimedes (287–212 BC), the great mathematician of antiquity, has long been known for ingenious discoveries and mechanical inventions. In fact those discoveries and inventions followed from his advances in geometry. For example, he was asked by Hieron, the Tyrant of Syracuse, to figure out if a crown was gold or whether the goldsmith had cheated the Tyrant by mixing silver with gold. Because gold is more dense than silver, the solution, Archimedes knew, was in finding the density of the crown. Density is the ratio of the weight of an object to its volume. The weight of the crown was easily determined by measuring it on a scale. However, the crown was an irregular shape. How could its volume be determined without melting it down?

After puzzling over the problem for some days Archimedes visited the baths and noticed that, as he entered the bath, some water spilled over the edge. It then came to him that his body, the crown, or any other solid object would displace exactly its volume on being immersed. The story concludes with Archimedes running naked to his home shouting, "Eureka! I have it."

Regarding the laws of levers, Archimedes is said to have asserted to Hieron, "If I had a lever long enough and a place to stand, I could move the earth" (Heath 1897). In fact, Archimedes worked out the laws of levers with such precision that applications of them played an important part in thwarting, for a time, the Romans who besieged Syracuse. Those applications, like any application of theory, were tests of it. The experiments below are designed as tests of his theory.

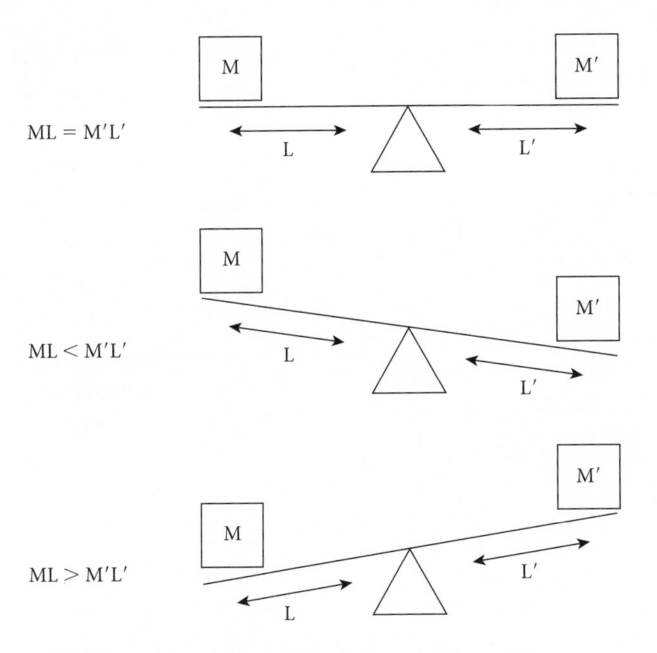

FIGURE 4.1. Models of Archimedes' Law of Levers

Figure 4.1 shows (1) laws of levers to the left and (2) the models to which the laws apply to the right. Each of the three models represents a beam pivoted at the apex of the triangle-shaped fulcrum. Two weights, M and M′, are placed on the beam at distances L and L′ respectively from the fulcrum.[4] For all three models, L = L′: the two distances from the fulcrum to the weights are the same. The three models represent three states of the system. In the first model, the two weights are equal and, because the Ls are equal, it follows that ML = M′L′. The empirical claim of the law of levers is that, when ML = M′L′, the system is balanced as shown in the model. In the second model, the weight to the left is less than the weight to the right. Because M < M′ and the Ls are equal, it follows that ML < M′L′. Now the empirical claim is that the system is imbalanced to the right. The third model reverses the second. Now M > M′ and, with the Ls still equal, it follows that ML > M′L′ and the system is imbalanced to the left.

Theory designs experiments. Therefore, an experiment will look like—and is predicted to behave like—the models in Figure 4.1. When an experimental setup is built like the theoretical model for it, Freese and Sell (1980) call the setup a *replica* of the model. If our replica is to behave as modeled, one need not be a rocket scientist to infer that it should have two ideal qualities: a frictionless fulcrum and a weightless beam. If the beam is weightless, its weight will not interact with the weights of M and M′. If the fulcrum is frictionless, the beam's attitude—whether it is horizontal or sloping in one or the other direction—will reflect the forces of M and M′. Unfortunately, the world does not contain weightless beams and frictionless fulcrums. What is to be done?

The answer lies in using the law of levers to infer how the experiment will be threatened by friction and beam weight. When friction is great enough, the system will remain in its initial state, balanced or imbalanced, although the weights are changed. If so, observations could wrongly falsify the law. If the law is correct, however, we can alter M or M′ to find the size of the friction force relative to the weights. For example, by adding small quantities of weight to M we can detect the threshold at which the beam will move and thus the size of the friction force. If we can find the size of the friction force, we can, by trial and error, find ways to make it smaller. Said differently, because we cannot create a frictionless fulcrum, we will design the experimental apparatus with friction force very low compared to the sizes of M and M′.

Regarding the beam, the law of levers asserts that, if one side of the beam is heavier than the other, the experiment testing $ML = M'L'$ will assuredly be affected. The system will behave as if $ML > M'L'$ or as if $ML < M'L'$. Fortunately, the system can be tested before the experiment. First, measure to be sure that $L = L'$. Then remove M and M′ and check for balance. If the beam balances, beam weight is equally distributed and theory asserts that the experiment will not be affected. The experiment will not be affected because adding a constant to M and M′ does not affect either the equality or the inequalities.

With the model effectively replicated, it remains only to set the remaining initial conditions and run the three experiments. We have already determined that $L = L'$. For the first experiment we need to set values for Ms such that $M = M'$ while the second experiment sets $M < M'$ and the third $M > M'$. To set these values we will not use a balance scale. Because a balance scale is based on the law of levers, its use—and our faith in it—depends on confirmation of the laws being tested. Instead, M values will be set using a spring scale. To run the experiments, M and M′ are placed on the beam and the resulting balance or imbalances observed.

As explained above, the models on the right of Figure 4.1 give the designs for the experimental apparatus. The laws to be tested are the expressions on the left that describe relations between ML values. The three attitudes of the beam shown in the figure are the predicted results. When experimental outcomes satisfy the models, the theory is supported. Further applications of the theory are not limited to cases like those tested, but, when they are, we will be more confident in our predictions.

We experience for the first time an important quality of theory-driven experiments. The results of the experiment are not surprising—at least, not to the experimenter. Nor should they be. The purpose of theory-driven experiments is not to make surprising discoveries unanticipated by the theory. Instead, theory anticipates the "discoveries" before the experiment. The purpose of experiments is to learn whether the theory's discoveries are sound. If we lived in a perfect world, no experiment should ever be conducted if its results were in doubt. However, theories and the people who apply them are fallible. Theory-driven experiments *do* make discoveries—at times to the delight and sometimes to the chagrin of the experimenter.

Criteria for Theoretically Driven Experiments

The principles of theory-driven experiments follow directly from the logic of the example above. We state them as five maxims:

1. Derive one or more models from the theory to be tested.
2. Use the theory to generate predictions by linking initial conditions to end conditions.
3. Build replicas, set initial conditions, and observe the end conditions.
4. Compare results to predictions and decide whether the theory is supported.
5. Make inferences from theory with greatest confidence to instances most theoretically similar to experiments supporting the theory—but predictions are not formally limited by that similarity.

The unifying idea behind these maxims is that the theoretic model or models and the corresponding experimental replica or replicas should be as similar as possible to each other. The experiment is a test of that similarity. A test succeeds when the model and its replica are initially similar, *and* similar at the conclusion of the experiment such that the model predicts the replica. The theory may also predict process. If so, processes linking initial and end conditions of model and replica may also be compared.

When two or more experiments are developed, as in the laws of levers, theory-driven experiments will take on a superficial similarity to difference experiments. That is, those schooled only in difference experiments will understand the three lever experiments as seeking to show empirical differences. Nothing could be farther from the truth. In fact, each of the three lever experiments could stand alone for the following reasons.

All five criteria can be satisfied by a single experiment. For example, setting ML = M′L′ and observing that the beam balances *as predicted* is a test of the theory. It is a test because *theory-driven experiments are concerned only with testing the correspondence between theoretical model and experimental replica—between prediction from theory and experimental results.* Theory-driven experiments are not concerned with empirical differences. Why then do many applications of theory employ multiple experiments that differ in only one regard from each other as did the three lever experiments? Is the aim to increase the number of confirming instances and thus reduce the likelihood that results are due to chance? The answers to these questions are found in a theory's scope of application, precise measurement of its concepts, and how effectively it designs simple models of complex phenomena.

The main purpose of multiple experiments is to investigate the scope over which the theory can be fruitfully applied. Consider the laws of levers. Now, suppose the theory's scope was limited just to those instances for which ML = M′L′. A theory that is known to be supported only when ML = M′L′ would have limited utility. A theory that is effective across the three cases of Figure 4.1 has much more utility. Of course, the scope of the theory of levers is not exhausted by the three experiments just discussed. Having applied the theory to not one but three cases, however, we will have greater confidence when it is applied in further experiments, even when they are not similar to the three of Figure 4.1.

The Moore experiments described below further illustrate why some theory-driven experiments use designs that mimic difference experiments. In some cases, measures of the theory's concepts are relatively inexact. Experiments built from the models Moore tested predict ordinal-level relations between status and influence. Without point predictions, some implications of the theory require multiple experiments to test them. As a result, Moore ran several experiments that permitted him to compare across experiments and determine that the theory's predictions were supported. Finally, some designs simplify complex phenomena so effectively that multiple experiments must be run to recapture complexity. Each of Moore's experiments focused on only one party to status-differentiated interaction (either high or low status subjects). Multiple experiments, some with high, others with low status actors, are required to compare findings with predictions for interaction between persons of unequal status.

Replication is also important because the results of a single experiment could be accidental. By accidental, we mean that some conditions not recognized by the theory produced the predicted results or made production of them impossible. Fortunately, the likelihood that results are accidental decreases rapidly as tests accumulate. Therefore, confidence in a theory sharply increases as the number of experimental runs increases. The best tests for cumulation of results are those in which initial conditions are varied one at a time such that, beyond accumulation, the scope of the theory is systematically investigated.

In concluding this discussion of the principles of theory-driven experimentation, we turn to the fifth maxim. It asserts that we are most confident when predicting instances theoretically similar to those investigated experimentally. However, the universe of predictions is not limited by that similarity. Those who have little experience working with theory find it surprising that the applications of theory are not limited to the range of phenomena over which it has been tested. On further consideration, much of the value of theory would be lost if applications were so limited.

Theory offers the possibility of extending our knowledge beyond what we know to what we seek to know. Among the reasons that we build theory, this is perhaps the single most important one. Nevertheless, to some, applying theory beyond the range within which it has been tested or outside its scope appears dangerously risky. But what is the alternative? The alternative is to proceed blindly without theory. The better alternative is to rely on tested theory. A theory that has survived many tests can be usefully extended to new situations even if those situations lie outside its scope or the range of previously successful tests. As a theory's applications are extended, each successful new application, beyond being an explanation or prediction, reflects back on the theory as a new test. Thus, does scientific knowledge cumulate.

Applying the Maxims to a Sociological Experiment

A great problem, known since antiquity (Aristotle [347 BC] 1962) and central to classical social theorists like Marx, Simmel, and Weber is how social structure affects human activity. We use Network Exchange Theory (Willer 1999), to explore the relationship between exchange structures and action in mixed-motive bargaining situations. Network Exchange Theory (hereafter, NET), uses the principle of rationality—all social actors seek to maximize their expected preference state alteration (Willer 1999:30)—to generate the models of Figure 4.2. The principle implies that (1) people have ordered preferences, and (2) their actions are designed to achieve the highest ranked of those ordered states, such that (3) their behaviors can be inferred in the models we draw. We begin with a mixed-motive social relation linking A and B. Then we show how A's and B's actions in that relation are affected by structural differences in the exclusive alternatives available to A or B.

Mixed-motive relations are those in which both actors are driven by two contrasting motives. When A and B can reach many agreements (1) any agreement that is better for A is worse for B and vice versa, and (2) there are agreements better for both A and B than confrontation (i.e., when they do not agree). Therefore, actors compete because agreements better for one are worse for the other. And actors cooperate because there are agreements that are better for both than failure to agree. Three models are shown in Figure 4.2. Each is composed of the mixed-motive A-B relation, which has a pool of 10 valued resources and two alternative payoffs external to and exclusive of the relationship. By "exclusive of" for A we mean that A can reach an agreement with B or accept the alternative payoff (shown by the arrow to the left of A) but not both—and similarly for B. With these ideas in hand, we can use the models to build experimental replicas.

In all three models if A and B agree on a division they will divide the pool of 10 valued resources. However, they will gain no resources in the relation if they fail to agree on a resource division. The principle of rationality implies that A and B prefer more to fewer resources and that neither will agree to a division in which it gains nothing. Next, we stipulate that resources are divisible in units of one. It follows that, in the first of the models, where both A and B have zero alternative payoffs: there is a range of nine possible agreements with a range of 9–1 preferred by A to 1–9 preferred by B. For any agreement, if A gains more, B gains less and vice versa. All agreements within the range are better than no agreement where A and B gain zero. Therefore, the first resource pool relation is mixed motive. The other two relations are also mixed motive but, as will be seen, the range of agreements is truncated by the alternatives.

Predicting payoffs for the models to the right of Figure 4.2 proceeds as follows. The first modeled structure is symmetrical. Furthermore, the two actors are identical because they are defined only by the principle of rationality. It follows that their payoffs (P_X) must be the same: therefore, $P_A = P_B$. A and B

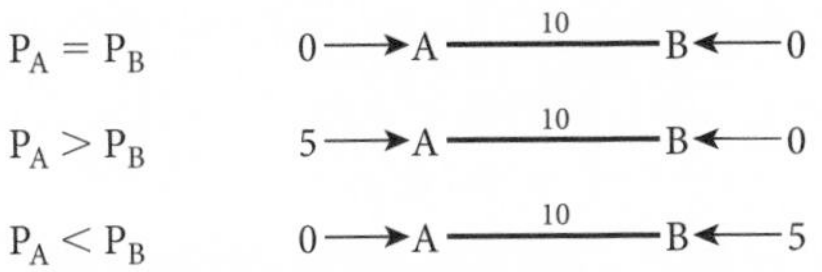

$P_A = P_B$ 0 ⟶ A —10— B ⟵ 0

$P_A > P_B$ 5 ⟶ A —10— B ⟵ 0

$P_A < P_B$ 0 ⟶ A —10— B ⟵ 5

FIGURE 4.2. Models of Symmetrical and Asymmetrical Exchange Structures

each gain 5. In the second model, A's exclusive alternative is 5 but B's is zero. According to the rationality principle, A will not accept any agreement with B for less than 5. Offering only 5 is risky for B because A is indifferent to B's offer and the alternative 5 and may choose the latter excluding B from any payoff at all. More precisely, with two identical alternatives, a purely self-interested A will chose randomly and thus exclude B half the time. To be sure of a non-zero payoff, B must minimally offer 6 and, because there are 10 to divide, B's payoff cannot be greater than 4. Therefore, $P_A > P_B$. The third model simply reverses the second. Because B now has the only nonzero alternative, to avoid a zero payoff, A must offer 6 or more, accepting 4 or less. Therefore, $P_A < P_B$.

Theory designs experiments. Therefore, an experiment will be designed to look like—and is predicted to behave like—the models in Figure 4.2. The experimental set up should be a *replica* of the model. If our replica is to behave as modeled, one need not be Max Weber to infer that it should have two ideal qualities. The experimental system should (1) be perfectly isolated with no payoffs directed from outside the modeled system and (2) have no "side" payoffs introduced by the subjects who we place in the A and B positions. Side payoffs are inducements that subjects make to others to get them to accept their offers.

Outside payoffs and side payments must be controlled because neither is part of the model. If there are no outside payoffs and A and B cannot introduce side payments, both are limited, as in the model, to dividing the 10 resources or to accepting the alternative payoff. The world does not contain perfectly isolated systems nor will rational actors fail to introduce side payoffs to better their agreements when they can do so. What is to be done?

The answer lies in using the rationality principle to infer how the experiment will be threatened by payoffs not already modeled. The rationality principle implies that activity can be predicted only when all relevant payoffs are known. For example, if a payoff outside the model is great enough compared with the ones modeled, it will be accepted and A and B will not agree nor will either accept the modeled alternative. Yet the rationality principle implies that, in all three models, A and B will exchange. If they do not, the absence of agreement might be taken as reason to call the principle of rationality into question. Thus, application of the rationality principle implies that tests of the models require the researcher to create systems as perfectly isolated from outside payoffs as possible.

How can the best replica be built? To eliminate side payoffs that A or B might introduce, the investigator could strictly monitor their communications and ensure that subjects never meet face to face. As found in Chapter 3, however, having experimenters directly interacting with subjects is not desirable. Therefore, a better alternative to having the experimenter monitor interactions is to create a setting in which all interactions are mediated though a computer interface. That computer interface limits subject communications to offers and agreements.

In principle, there is no reason why social science experiments cannot be as precise, if not more precise, than physical science experiments. For the test of the law of levers it was not possible to create a frictionless fulcrum. By contrast, the experimental design just described could be called "frictionless" in that its replicas precisely correspond to its models. Arguably, these designs replicate their models more accurately than the designs used to test the law of levers.

Having built the computer interface, it remains to motivate subjects consistent with the rationality principle and run the experiments. To motivate them to prefer higher to lower payoffs, subjects are paid by points earned. Add to that condition the experimental demand that they should seek their best payoff. In running the experiments, subjects are placed in separate rooms and each subject negotiates from either the A or the B position. (To avoid a possible bias, the PC system allows both subjects to see themselves in the A position while their partner is B.) Each design was run for 10 trials and experimental outcomes corresponded to predictions drawn from the models. In the first experiment, where the prediction was $P_A = P_B$, A and B both averaged 5 points. In the second experiment, where $P_A > P_B$ was predicted, A gained more than B. In the third, also as predicted, B's payoffs were greater than A's.

For the second time, we find that experimental results are not surprising—nor should they be. The predictions were known before the experiments were run. Thus, unless the theory was false or the experimental design was faulty, the outcomes should have been the ones predicted. The experiments produced no new discoveries and none were expected. Here, as in all science, discovery is important, but, as in all theory-driven experimentation, the discovery was the models' predictions.

The purpose of the experiments was to test whether the rationality principle could be successfully applied as in the models drawn in Figure 4.2. Now, we look back at the experimental designs and comment on their fit with the five maxims of good design. The maxims are given in italics:

1. *Derive one or more models from the theory to be tested.* Three models were built, each of which corresponds to a model shown in Figure 4.2.
2. *Use the theory to generate predictions by linking initial conditions to end conditions.* The models gave the initial conditions and three predictions were generated using the rationality principle. They are shown to the left of the figures.
3. *Build replicas, set initial conditions, and observe the end conditions.* The computer mediated setup allowed the construction of replicas very nearly identical to their models. The experimental runs began by setting conditions corresponding to the models. They were maintained during the experiments and, at their conclusion, A's and B's payoffs were observed.
4. *Compare results to predictions and decide whether the theory is supported.* Results for each experiment corresponded well to predictions. Application of the rationality principle to the models was supported.
5. *Make inferences from theory with greatest confidence to instances most theoretically similar to experiments supporting the theory—but predictions are not formally limited by that similarity.*

Again the unifying idea is that each model and its replica be similar initially and have similar end conditions. The test of the theory is that similarity. Here the test supported the theory. The theory was supported because each replica and its model were initially similar and each had similar results. Those who have yet to see the distinction between empirically driven and theoretically driven experiments would be inclined to test whether results for the three replicas are different from each other. Although there is no harm in such a test, its results are quite irrelevant to the experiment as a test of theory.

Regarding the fifth maxim, the rationality principle could be applied to many models that are unlike the three just studied. For example, there is a structure like the second in Figure 4.2 but in it a second 10-resource relation with C is substituted for the fixed alternative. Now A has two exclusive alternatives, to exchange with B or with C, but not both. The rationality principle predicts a bidding war between B and the newly arrived C. To reach an agreement, C would have to outbid B, and, with that bid, B would have to outbid C and similarly. No matter what the first bid is, the end point of the bidding process is the 9–1 offer favoring A. Thus, the rationality principle predicts the 9–1 resource division, but it cannot predict which of B or C succeeds in exchanging with A. This application is a straightforward extension of the second experiment. Thus, we should be confident of the prediction.

Another step removed is a model that allows A to send a costly negative sanction to B. Over repeated trials, a rational A will send that sanction if B does not agree to the offer A wants. Add that negative sanction to the first of the models in Figure 4.2. The model is no longer symmetrical. A is advantaged, and the rationality principle predicts $P_A > P_B$. The new structure is two steps removed from the research already completed, and we have less confidence in predicting the effect of adding negative sanctions than in predicting resource divisions when A has two exclusive resource pool relations. Nevertheless, using the theory for this new prediction is far better than making "educated guesses" not grounded in theory.

Falling Bodies and Status-Based Influence

The previous section identified five principles for theory-driven experiments and stated them as maxims. Inferred from experiments testing the laws of levers, the principles were applied to sociological experiments that used three simple structural models to test the rationality principle. In this section we offer two more example experiments and evaluate them according to the five maxims. The first is from physics, Galileo's studies of falling bodies, while the second is Moore's study of the effects of status. Because both are theory-driven experiments, if they are well designed, they should be similar to each other, at least insofar as both satisfy the five maxims. There will be further similarities.

The Study of Falling Bodies

The problem of falling bodies is the problem of objects moving freely under the influence of gravity. It is a problem that has puzzled scientists since antiquity. Aristotle's theory was concerned only with the speed of fall, which he claimed varies with a body's *natural rate of fall*. In his theory, natural rate of fall is an innate property of a body that is (1) directly proportional to its weight and (2) inversely proportional to the density of the medium through which it falls ([330 BC] 1961:162). His theory was widely accepted until Galileo.

Galileo ([1636] 1954) dispensed with Aristotle's theory in two ways. First, he used Aristotle's theory to design an experiment. Galileo dropped a 200-pound cannon ball and a musket ball weighing less than .5 pound from 300 feet. Applying Aristotle's theory, the cannon ball should have fallen 400 times faster than the musket ball (400 = 200 ÷ .5), but it did not. Galileo observed that the cannon ball reached the ground less than a foot ahead of the musket ball (Galileo [1636] 1954:62).[5] He had shown that Aristotle's theory was empirically wrong but the work of falsification had only begun.

Galileo initiated a second phase of research in which he examined Aristotle's theory to show that it is not logically consistent. To show that it was not, he offered the following *thought experiment*. Begin with Aristotle's claim that heavy bodies naturally fall faster than lighter bodies. Next, make a composite body with the lighter one on top. The lighter body pulling up ought to retard the fall of the heavier body while the heavier body pulling down should increase the speed of the lighter body. Thus Aristotle's theory predicts that the composite body falls at a rate between the rates of the two bodies. Nevertheless, that prediction cannot be true because the composite body is heavier than either of its parts and must fall faster. Therefore, Aristotle's theory makes contradictory predictions. Galileo had shown that Aristotle's theory is empirically wrong and that it is logically inconsistent. Yet to falsify Aristotle, it was necessary for Galileo to offer a better theory.

The first step in offering a new theory was to shift the focus of concern from natural speed, where it had remained for two millennia, to *natural acceleration*. Natural acceleration is the rate of *change* in the speed of a falling body. Galileo's theory is that all falling bodies accelerate at the same rate. To generate predictions, he proposed the concept of *uniformly accelerated motion*. "A motion is said to be uniformly accelerated, when starting from rest, it acquires, during equal time intervals, equal intervals of speed" (Galileo [1636] 1954:162). As an example of uniformly accelerated motion, when x is distance, at the end of one second, a body will be falling at a speed of x/second, at the end of two seconds at the speed of $2x$/second, at the end of 3 seconds at $3x$/second, and so on. For example, today we know that the force of gravity on Earth is approximately 32ft/sec^2. Following Galileo, after the first second a body will be falling at a speed of 32ft/sec, after the second, 64ft/sec, after the third, 96ft/sec, and so on.

Recall from previous examples that experimentalists frequently devise new measures to study their phenomena. Galileo faced a series of measurement problems as he tried to build an experiment to test his theory. The simplest test would have been to measure the speed at which a body was falling at any given instant in time. Was it falling at $3x$/second after 3 seconds of fall? To answer that question, however, he would have had to measure instantaneous speed—that is, speed at an instant in time. That sort of measurement was not possible until the development of radar (and later lasers) in the twentieth century. On the other hand, distance could be readily measured, much as we do now, by placing a standard x times along a straight line. Thus, before experiments could test his theory, it was necessary to extend the idea of uniformly accelerated motion to relate time to distance fallen.

Galileo reasoned as follows: Because acceleration is uniform, in the interval t_0 to t_1, average speed will be the average of zero speed at the beginning of the interval and x/second at the end. Similarly for the interval t_1 to t_2 average speed will be the average of x/second and $2x$/second. By extension, when V_i is the average speed for an interval, the average speeds for the intervals t_0 to t_1 and t_1 to t_2 are

$$V_1 = \frac{0/1 + x/1}{2} = \frac{x}{2};\ V_2 = \frac{x/1 + 2x/1}{2} = \frac{3x}{2} \quad (1)$$

each in distance per second. The distance traveled by the end of the first second is $x/2$ and by the end of the second is $x/2 + 3x/2 = 2x$ and similarly for each additional second. More generally, where t is time, Galileo determined that $x \propto t^2$. (Distance fallen is proportional to time squared.)

All that remained for Galileo was to measure time but measuring time also posed a serious problem. All mechanical clocks, including those built by Huygens, are based on Galileo's theory of pendulum motion. Huygens would not invent the first mechanical clock until more than a century after Galileo's writings. Lacking a mechanical clock, Galileo devised an alternative solution:

> For the measurement of time, we employed a large vessel of water placed in an elevated position; to the bottom of this vessel was soldered a pipe of small diameter giving a thin jet of water, which we collected, the water thus collected was weighed. (Galileo [1636] 1954:179)

Galileo's solution depended on the fact that each time interval is proportional to the weight of water collected. A water clock could be reasonably precise, but not for very short periods of time. How could a water clock be used to time rapidly falling bodies precisely? Because a better clock could not be devised, the phenomenon would have to be slowed. Freely falling bodies move too fast, but the acceleration of a body down an inclined plane can be made slower or faster by adjusting the angle of the plane from vertical. To that end, Galileo devised a wooden molding with a polished channel down which a round bronze ball was rolled. He reported his results in the following way:

> Next we tried other distances, comparing the time for the whole length with that for the half, or with that for two-thirds, or three-fourths, or indeed for any fraction; in such experiments, repeated a full hundred times, we always found that the spaces traversed were to each other as the squares of the times, and this was true for all inclinations [angles] of the plane. (Galileo [1636] 1954:179)

As we might say today, the series of experiments offered strong support for the prediction from Galileo's theory that falling bodies are uniformly accelerated.

Now we apply the five maxims to Galileo's experiments. The maxims are given in italics:

1. *Derive one or more models from the theory to be tested.* Galileo devised a model for the rate at which falling bodies fell based on the ideas of uniformly accelerated motion, distance, and time.
2. *Use the theory to generate predictions by linking initial conditions to end conditions.* Today we would run a few experiments to derive g, the gravitational constant, and then use it in the equation $x = 1/2gt^2$ to predict results from the remaining experiments. Galileo did not derive g; instead, he used the theory to predict results for iterations of the basic model. Because his theory claimed that distance varied with time squared, he made predictions for falling balls that traveled different distances.
3. *Build replicas, set initial conditions, and observe the end conditions.* Galileo's replica was constrained by his method of measuring time. To slow the process, the design represented, not freely falling bodies, but bodies falling down an inclined plane. The plane was made as

flat as possible so that, according to his theory, acceleration could be uniform. The initial conditions were a virtually frictionless plane, length of the plane, and the angle of inclination.

4. *Compare results to predictions and decide whether the theory is supported.* Results corresponded to predictions in the following sense. Consistent with his law, for every set of initial conditions, each observed distance traveled was proportional to time squared.
5. *Make inferences from theory with greatest confidence to instances most theoretically similar to experiments supporting the theory—but predictions are not formally limited by that similarity.*

Considering the fifth maxim, an important series of studies extended these experiments. Galileo was quite ready to infer from his inclined plane studies to freely falling bodies. That inference was strengthened by showing that the time of fall of a body increased with the inclination of the plane. Certainly, had his studies not been limited by his ability to measure small increments of time, he could have shown that the rate of acceleration on his plane approached that of free fall as it approached the vertical. In any case, the next step of his research was to predict the path of projectiles. To do so, he treated their motion as composed of two parts, one vertical that was uniformly accelerated (for the projectile was freely falling) and one horizontal where velocity was constant. Taken together the two motions defined a parabola—or a series of parabolas for a series of projectiles. The next step, taken later by Newton, was to link these extensions to Kepler's theory of planetary motion, but that is another story. (See Willer 1987:12.)[6]

Experiments on Status-based Influence

Social influence processes have been observed since antiquity. Remember that Aristotle thought of falling bodies as having natural speed of fall, an innate quality. He also attributed influence to the influenced person's greater or lesser susceptibility, again an innate quality (Aristotle [347 BC] 1962:199ff). Susceptibility, he asserted, is a constant or natural quality for each individual. If so, a person's susceptibility to influence will be constant across social situations.

Modern social scientists question the Aristotelian position. From the Asch experiments, we have seen that conformity varied with the number of confederates. Technically, Asch's results do not falsify Aristotle because, as shown in Chapter 3, conformity was due to power not influence. The experiment to be analyzed now does falsify Aristotle. The theory on which it rests grew out of the Bales experiments discussed in Chapter 2 and it is about influence. Bales' studies (1950, 1999) found that the extent to which one person could get others to change their beliefs and thus their behavior varied with their respective positions in social structures. Contrary to Aristotle, the extent to which A is influenced by B is not constant across social settings, but varies with B's status. More generally, Marx, Weber, and Simmel devised the modern conception of a social relationship as a situation in which each person orients her or his behavior to the actions of others in the situation (cf. Weber [1918] 1968:26). Their ideas about how social relations affect changes in behavior are embedded in contemporary conceptions of power and influence.

Here we revisit the modern distinction between influence and power, both of which occur in social relations. For influence, behavior change occurs because beliefs have changed. When B changes her beliefs due to interaction with A, and, due to those changed beliefs, B alters her behavior, B's behavior was influenced by A. Power affects behavior through sanctions, not belief change. Power occurs when A changes B's behavior as a consequence of offers to send positive sanctions or threats to send negative sanctions. Power exercise is through the imposition of positive and negative sanctions, but sanctions are not part of the influence process.

Nevertheless, recent experimental research has shown that influence can also produce power events (Thye, Willer, and Markovsky 2006). Because influence and power are undoubtedly related in a number of ways, only some of which are now known, any study of influence must control for power relations and eliminate them as far as possible.

Many studies of social influence have focused on behavior in task groups. Task groups are organized to complete one or more tasks. Most committees are task groups. Extensive research has found that those task group members who are identified with specific social categories (e.g., males or Caucasians) have more influence than those who are members of other social categories (e.g., females or blacks). For example, mixed-gender juries more often choose foremen who are white, male, and have professional occupations (Strodtbeck, James, and Hawkins 1957). Patterns like that were not well understood before the development of Status Characteristics Theory (Berger, Cohen, and Zelditch 1966, 1972; Berger, Fisek, Norman, and Zelditch 1977). (From now on we use SCT to refer to Status Characteristics Theory.) We will now identify basic SCT concepts and show how the theory is used to design experiments and generate hypotheses.

Joseph Berger and his colleagues developed SCT. They recognized that attributes like gender and occupational categories are examples of *status characteristics*, an idea introduced by Hughes (1945). Next, they refined the idea by identifying and defining *specific* and *diffuse* status characteristics. Specific status characteristics, like musical ability, are attributes or personal qualities for which (1) the states (e.g., high and low ability) have different social value or rank, and (2) each state is linked to a specific performance expectation state that has the same sign as the ranked state. For example, those with high musical ability are *expected* to sing better or play an instrument better than those with low musical ability.

Diffuse status characteristics are those, like race or gender, for which (1) the states (e.g., female and male) are ranked, (2) each state is linked to states of one or more specific characteristics (females are expected to be more nurturing but less gifted mathematically than males), and (3) each state of the diffuse characteristic is linked to a *general performance expectation state* that has the same sign as the state in question. As an example of a general performance expectation state, in contemporary U.S. society, females are generally expected to be less competent than males.[7]

SCT explains why the influence rankings of group members in task groups (e.g., juries or work groups) are related to societal rankings of states of the status characteristics they possess. The theory's scope of application is governed by a set of explicitly defined scope statements (SCs).

SC1. Two actors (P and O) are required to perform a valued collective task, T.

SC2. Both P and O believe that success at T depends on having a specific task-related ability.

SC3. Neither P nor O knows how much of the specific task ability either of them has.

SC4. P and O possess different states of only one diffuse status characteristic.

The first statement has two parts: (1) task outcomes must be valued in the sense that group members believe that it is possible to succeed or fail; and (2) the task is collective requiring group members to work together (cooperate) to complete it. The second scope statement is designed to eliminate situations in which group members believe that skill is unnecessary or irrelevant to task success (e.g., success is a matter of luck). The third statement requires group members to be ignorant of their skills. If they are aware of their task-relevant skills, they have no need to use other information (e.g., status information) in the task situation. Finally, SC4 establishes the simplest status situation possible.[8]

SCT offers four theoretical statements (TSs), or arguments, that explain why rankings of status characteristics and influence rankings are related, even in situations where the characteristic initially seems irrelevant to the task.

> TS1. If group members have different states of a status characteristic or if the characteristics they posses are connected to the task, they will activate (pay attention to and use information about) the states of the characteristics they possess.
>
> TS2. If members consider their status rankings, they will believe that the states of those characteristics are connected to states of the ability that determines success or failure at the task.
>
> TS3. If members believe that status rankings are connected to the task ability, they will expect that states of the ability are distributed exactly as their status rankings. That is, those with high status are expected to have high states of the task ability.
>
> TS4. Members who are expected to have high states of the task ability exert more influence than those who are expected to possess lower states of the task ability.

A test of SCT must take into account the fact that power and influence processes can have similar effects in social relations. Much as Galileo used smooth inclined planes to reduce friction in his studies of falling bodies, researchers studying influence must exclude power from the experimental situation. Power can be excluded from an influence experiment by eliminating opposing interests among those engaged at a task, or by eliminating the possibility that subjects can sanction each other, or both.

James C. Moore (1968) used SCT to design an experiment in which female subjects believed that they were interacting with another female subject. He gave them information that they and a fictive partner possessed different states of a diffuse characteristic (educational status). Next, Moore created a task that required subjects to judge whether more of the squares that formed a grid in a rectangular box were shaded than not. Figure 4.3 is one of many slides subjects viewed. The experimenter told subjects that they were to make

FIGURE 4.3. Contrast Sensitivity Slide for SCT Experiments

as many correct judgments as possible and to use any available information to help them make correct judgments.

This judgment task was designed to study influence where there is a change of belief, not power-based conformity where there is none. Looking critically at Figure 4.3, it may seem difficult to make an accurate judgment. In fact, the judgment task was purposefully ambiguous, for each box viewed by the subjects had between 55 and 45 percent black (shaded) squares. The task is designed to appear difficult, not impossible. Thus, when subjects' responses indicate a change from one judgment to another, that change should reflect a change of belief. Post-experimental interviews support the inference that subject's beliefs do change. By contrast, in the Asch experiment where differences of line lengths are obvious, changes in subjects' responses were not due to changed beliefs. Thus, Asch studied conformity due to power, but Moore studied influence. We turn now to how influence was exercised.

The instructions before the experiment were intended to eliminate conflicts of interest and power processes. Subjects were assured that the experimental task was cooperative not competitive. Subjects did not encounter other subjects before the experiment and could not see or hear their partners during the experiment. Subjects were also assured that they would not meet their partners after the experiment. These assurances were intended to eliminate the possibility that a subject's behavior would be affected by the anticipation of power processes like those Asch studied.

The experiment began when the experimenter explained that the study was split into two phases so that he could learn which of two methods permitted subjects to make more accurate judgments. Subjects would be given scores based on how many accurate judgments they made. Pairs of subjects (one real, one fictive) were seated in separate rooms. Each phase consisted of many trials. On each trial in the first phase, the subject viewed a slide for 5 seconds. Then the subject pushed either a "black" (shaded) or "white" button to record her judgment. In the second phase, subjects were given more time and more information as they processed 40 slides. Each saw a slide and made a judgment (now called *initial choice*) as in the first phase. Immediately after the initial choice, each subject saw her partner's choice and then viewed the slide for another 5 seconds. At that point, the subject pushed a black or white button and recorded her final choice.

The experimental instrumentation had two components. The first was a slide projector and slides that were viewed as described. Second, experimenters controlled the information the subject received by filtering initial responses through an *Interaction Control Machine* (ICOM) devised by Joseph Berger and his colleagues (Berger, Cohen, and Zelditch 1972:249). Viewing lights turned on and off by switches, the experimenter saw the subject's initial choices. Then the experimenter pushed a button on a master control panel of the ICOM to switch on a light on the subject's panel indicating the choice of the fictive partner. Across trials the experimenter's feedback to the subject showed that the fictive partner *disagreed* on 28 of 40 initial choices and agreed on the remaining 12. Today experiments of this kind are run in effectively the same way but a PC replaces the slide projector and ICOM. The PC setting is described in Chapter 5.

How did Moore measure influence in this situation? Moore used a measure Berger and his colleagues devised to measure *resistance to influence* as follows. A subject is not influenced if she does not change her judgment to be the same as that reported for her fictive partner. For example, when a subject initially pushed the black button, saw that her partner had pushed the white button, and then pushed the black button for her final choice, she was not influenced by her partner in that trial. This pattern is labeled a *stay response. Only initial disagreement followed by changed judgment indicated influence.* Using the ICOM, Moore recorded the instances of stay and influenced responses. Next, he calculated the probability of the subject's stay response, p(S) using the formula

$$p(S) = \frac{N_S}{N_D} \tag{2}$$

where N_S is the frequency of stay responses and N_D is the number of trials on which the partners initially disagree. The p(S) is inversely related to influence. That is, people with lower p(S) values are influenced more often than those with higher values.

All subjects were students at a junior college. Half were told that their partners were from a local high school. That is, half the subjects were told that they had higher educational status than their fictive partners. The other half were told that their partners were from a local university thus placing subjects lower in educational status than their partners. These two groupings were subdivided into two additional subgroups. One subgroup got only the information about their partner's educational status. Subjects in the remaining subgroup were also told that educational status was positively related with task success. Thus, Moore created four initial conditions:

1. The subject was told that her partner is a high school student (Subject High Status, HS).
2. The subject was told that her partner is a high school student *and* that junior college students usually do better than high school students (Subject High Status, Explicit Relevance, HS-ER).
3. The subject was told that her partner is a university student (Subject Low Status, LS).
4. The subject was told that her partner is a university student *and* that junior college students usually do worse than university students (Subject Low Status, Explicit Relevance, LS-ER).

The theory predicts that high status persons are less susceptible to influence than low status persons, that is, they have higher p(S) values. That is what Moore found. High status subjects in conditions 1 and 2 had higher proportions of S responses than low status subjects in conditions 3 and 4 (HS = .696; LS = .613). The experiment supported SCT's prediction linking status to influence. Furthermore, Moore's experiment also permitted him to show that the link between status and influence is consistent with the *processes* described by the theory.

SCT asserts that, unless there is evidence to the contrary, people will use status information to infer task effectiveness. It follows that subjects who were *not* told about task relevance ought to use information about their relative status to infer that educational status is task relevant. Therefore, there should be no difference in p(S) values between those given relevance information and those not given that information. The theory was supported. Moore found that high status subjects had nearly identical p(S) values whether given relevance information or not (HS = .699; HS-ER = .693). Similarly, low status subjects' p(S) values were not significantly affected by relevance information (LS = .634; LS-ER = .593). Thus, Moore's findings support SCT's prediction that status affects influence and its prediction about *how* status affects influence.

We apply the five maxims to Moore's experiments. (As before, the maxims are in italics.)

1. *Derive one or more models from the theory to be tested.* Moore devised a two-stage decision-making model. In the first, a person makes independent decisions. In the second, a person decides, sees the decision of a real or fictive other—a source of influence—then makes a second or final decision.
2. *Use the theory to generate predictions by linking initial conditions to end conditions.* Moore used SCT to make four predictions that correspond to variations on two elements of the theory. First, information about relative status affects resistance to influence: (1) High status subjects resist more and (2) low status subjects resist less. Second, relevance (linking subjects' status to the task) does not affect resistance to influence: (1) High status subjects resist influence equally regardless of (explicit or implicit) relevance and (2) low status subjects resist influence equally regardless of relevance.
3. *Build replicas, set initial conditions, and observe the end conditions.* Subjects viewed slides like the one given in Figure 4.3 above and made initial and final judgments for 40 trials (12 agree and 28 disagree). Two measures were varied to set conditions for the predictions: (1) Subjects were told they were at a higher status (junior college vs. high school) or lower status (junior college vs. four-year college) than their partners and (2) were given no information about the relationship between educational status and task ability or given information that task ability corresponded to educational status (i.e., high education status meant high task ability). For disagreement trials, researchers observed initial judgments, final judgments, and whether changed judgments were consistent with the judgments of fictive partners.
4. *Compare results to predictions and decide whether the theory is supported.* Results were consistent with predictions in the following

sense. High status subjects had higher p(S) values than did low status subjects, indicating that they were influenced less. Also, task relevance had no effect on resistance for either high or low status subjects.

5. *Make inferences from theory with greatest confidence to instances most theoretically similar to experiments supporting the theory—but predictions are not formally limited by that similarity.*

Consistent with the fifth point, there is an important series of extensions based on these experiments. Several replications followed Moore's experiments. Some used different status characteristics like Air Force rank (captain, sergeant, airman third class) or scores on the Armed Forces Qualifications Test (AFQT). Researchers had confidence in predicting the results of these studies because their designs were similar to Moore's design. However, other experiments used two or more characteristics (an Air Force captain with high AFQT scores vs. a sergeant with low AFQT scores). The theory was supported when multiple status characteristics were consistent: that is to say, when two characteristics are both high for one person and both low for the other. (See Berger, Cohen, and Zelditch 1972 for descriptions of these studies.) However, studies that involve multiple characteristics raise an important question. What if status characteristics are inconsistent? To complete the scope extension, the theory needed to predict for two or more characteristics that were allocated *inconsistently* (e.g., high and low vs. low and high). Hughes (1945) had anticipated this problem decades earlier when he discussed inconsistent pairings of race and occupation as, for example, when a black (low) doctor (high) treated a white (high) patient (low).

The theorists devised two alternative hypotheses—the balancing and combining hypotheses. The balancing hypothesis proposed that group members "balanced" the situation by ignoring or discounting some status information. For example, a black doctor might ignore race status while his white patient could ignore occupational status. Alternatively, the combining hypothesis assumed that group members combined the information in a way akin to taking an average or weighted average of the information. Experiments were built to select between the hypotheses (Berger and Fisek 1970; Berger, Fisek, and Crosbie 1970). The results supported the combining hypothesis leading to a theoretical reformulation that included a combining mechanism. In this instance, designing experiments initially outside a theory's scope led to new discoveries in theory that were subsequently supported experimentally. (See Berger et al. 1977.)

Finally, we comment on Moore's use of multiple experiments. The theory predicts that high status actors will resist influence more than their low status partners. But subjects never interacted with real partners. Consequently, Moore ran multiple experiments with subjects in either high or low status roles and compared their results to determine if the theory's predictions were met. The apparent similarity to difference experiments is a function of theory-driven logic and Moore's tight experimental control, *not* the logic of difference experiments.

The Geometry of Experiments: Optics and Exchange

Because theory is the method of the sciences, when theories from different sciences have similar structures their experiments are designed similarly. Geometric Optics Theory in physics and Elementary Theory in sociology are two such theories. Both theories have principles and laws. A central principle in optics is the principle of the rectilinear propagation of light. Elementary Theory applies the principle of rationality. Both theories rely heavily on graphic representations to build models and their two principles play central roles in determining how the diagrams are drawn and interpreted. In fact, both theories employ three components, principles, laws, and diagrams, to generate predictions and to design experiments that test them.

Principles must be applied within the framework established by the theory—not in isolation. For example, the principle of the rectilinear propagation of light asserts that "light travels in straight lines," but reflection redirects light, and refraction bends light as does any gravitational field. Therefore, the principle of rectilinear propagation is false when applied in isolation or treated as an empirical generalization. It is not false when it is properly applied as one element of a theory of geometric optics. In that theory, the principle implies that, when composing optical diagrams, the theorist should represent light rays as straight lines. Then laws of reflection and refraction tell how to reorient the lines when they encounter a mirror or prism.

Similarly, Elementary Theory's rationality principle, that "all actors act to maximize their expected payoffs," should be applied only in the context of the theory, not in isolation. Applied alone it appears to assert that all humans are perfectly sane, every human activity is selfishly motivated, and activity is selected without regard to the actions of others. All are false generalizations. But the principle is not a generalization. Instead, it is a direction to the theorist concerning how preferences attributed to actors are linked to behavior. Nothing in the principle excludes altruistic preferences. More generally, in both these theories, and in other theories as well, principles are not assertions about the world. Instead, they are used with laws to create models for phenomena, generate predictions, and design experiments.

Applying a Theory of Geometric Optics

The study of geometric optics can be traced to Euclid around 280 BC. Euclid was the first to assert that light travels in straight lines. However, like other Greeks after Plato, Euclid had the direction of travel backward. For him light traveled from the eye to external objects and not the reverse as we today know to be the case. Euclid also correctly stated the law of reflection in his *Catoptics*, a law we apply below. The ancients were less successful with the law of refraction. First attempts by Ptolemy of Alexandria were finally overturned by Snell in 1621. Snell's Law was independently discovered by Descartes 16 years later. In fact, such independent discoveries are not unusual in the sciences.

We now explain how optics models are used to design experiments but we do not report who conducted the optics experiments. The law of reflection, like the laws of levers, is so old that saying who first tested it is impossible. In any case, the law is routinely tested by students in introductory physics classes using designs like the ones described here.

The law of reflection asserts that

$$i = r \tag{3}$$

which is read, the angle of incidence, i, equals the angle of reflection, r. The law generates a model like that represented in Figure 4.4. Figure 4.4 is a model for reflection of light from a plano (flat) mirror. Beginning from the left, a straight line represents the light ray traveling toward the horizontal mirror at the angle i from the normal. The normal, given as a dashed line, is an imaginary line at right angles (90°) to the mirror. To the right, a straight line at the angle r from the normal represents the reflected light ray traveling away from the mirror.

The model of Figure 4.4 is the design for an experimental replica for testing predictions from the reflection law. Initial conditions for the experiment are (1) a light ray source, (2) a flat reflective surface, and (3) the angle i. Angle r can be predicted by applying the law and the prediction can be compared with observations. Because i can take on any value $0° < i < 90°$ and similarly for r, the model serves as a design for many experiments for which i is varied

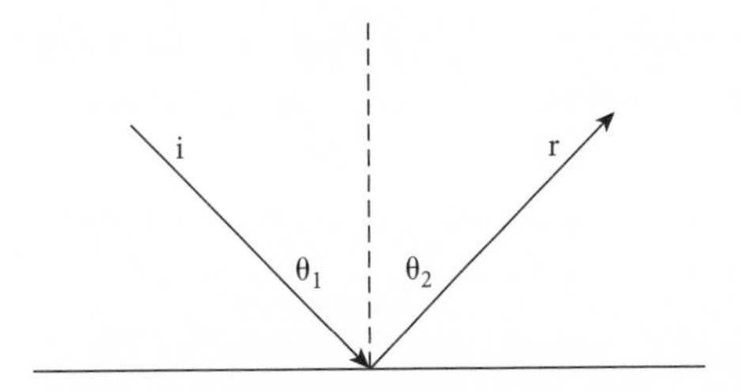

FIGURE 4.4. Reflection from a Flat Mirror

and compared with an observed *r*. Those who would like to carry out these experiments will find laser pointers to be excellent sources for light rays.

As shown in Figure 4.5, the theory of geometric optics finds the focal point of any curved mirror by a straightforward extension of the above. The model is built in the following way. First a curved mirror is drawn. Then, at the points A and B, two tangents to the curved mirror are drawn. For each, draw a normal line at 90° from the tangent. At A and B, two parallel light rays traveling from the left encounter the curved mirror. At A the ray encounters the mirror at the angle *i* and is predicted to be reflected at the angle *r* and similarly for the second light ray incident at B. For both, the prediction follows from the law, $i = r$. Then the focal point of the mirror is the point F at which the two light rays intersect.

The model makes it immediately obvious how to build an experimental replica to test it. Initial conditions for the experiment extend those of the previous experiments: Create (1) two light ray sources directed parallel to each other, (2) a curved reflective surface, and (3) the angles i_i. The law of reflection predicts the angles r_i and the focal point F, which can be compared with

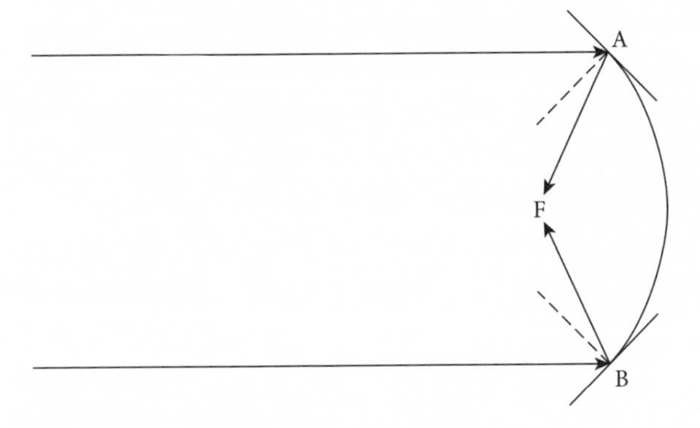

FIGURE 4.5. Reflection from a Curved Mirror

observations. (The focal point is the point at which the image reflected by the mirror will be seen.) Again the model is a design for a series of experiments, in this case for mirrors of diverse sizes and radii of curvature.

Now the five maxims are applied to the optics designs proposed above. The maxims are given in italics:

1. *Derive one or more models from the theory to be tested.* Two models are proposed above, one for flat mirrors and one for curved mirrors. In each case, the law of reflection and the principle of rectilinear propagation generate the model.
2. *Use the theory to generate predictions by linking initial conditions to end conditions.* Assigning varying values to *i* in both models, the law of reflection generates predictions for *r*.
3. *Build replicas, set initial conditions, and observe the end conditions.* For these experiments, build experimental replicas of plano and curved mirrors and set *i*, the angles of incidence of the light ray. The result to be observed is *r*, the ray's angle of reflection.
4. *Compare results to predictions and decide whether the theory is supported.* These experiments have been run many times and results routinely support the theory.
5. *Make inferences from theory with greatest confidence to instances most theoretically similar to experiments supporting the theory—but predictions are not formally limited by that similarity.*

Regarding the fifth point, one important extension was already included in the discussion, the extension from the plano mirror to the curved mirror. Adding the law of refraction, models can be generated to show how light rays bend as they pass through different media (e.g., from air through a glass lens to air). More complex models are built by combining simpler ones. Light rays can be traced through several lenses and mirrors to model telescopes of various designs. Finally, codifications of the theory have been offered. Fermat ([1662] 1896), the French mathematician, showed in 1658 that the principle of rectilinear propagation and the laws of reflection and refraction are all consistent with the idea that light travels the path that takes the least time.

Applying a Theory of Exchange Structures

It has been known at least since Marx ([1867] 1967) and Weber ([1918] 1968) that exchange can favor some actors over others. Exchange relations are symmetrical: they do not favor one actor over another. It follows that differences in benefit must stem, not from the exchange relation itself, but from the structure in which it and others like it are embedded. To explain power exercised in structures, Marx formulated the concept "separation." In the exchange structures of capitalism, workers are separated from the means of production because the means are owned by capitalists. Separation together with a reserve industrial army of the unemployed meant that workers' wages will move toward the minimum while capitalists' profits move toward the maximum.

Weber agreed that separation is an important structural power condition and subsequently used it to explain the centralization of power in bureaucracies where officials are separated from ownership of their positions. Recently, Corra (2005) explained that Marx and Weber's identification of separation as a structural power condition refers to exactly the same power condition as has been extensively investigated recently by Elementary Theory. That power condition is *exclusion* (Willer 1987, 1999). Actors who can be excluded from exchange relations are less powerful than actors who cannot be excluded.

The models and experiments discussed here investigate the impact of exclusion on exchange structures. The models will show that actors who are favored by exclusion will receive favorable exchange ratios that give them high payoffs at the expense of others who will receive unfavorable exchange ratios and low payoffs. We begin by showing how to model an exchange relation and then how to model structures with and without exclusion. The models built here are exactly the same as ones used by Brennan (1981) to build his pioneering exchange experiments.

The models in Figure 4.6 are networks that contain exchange relations. Each exchange relation is composed of two signed vectors, called sanctions. Sanctions are the acts sent and received by the connected actors and from which they gain payoffs. This discussion will focus on the A–B relation but because all relations are the same, it also describes the conditions of all the remaining relations.

In the A–B relation (e.g., Figure 4.6a), B has a single positive sanction that it can send to A. The payoff to A is 10 as indicated by the "+10" at the A end of the sanction. The resource has no value for B as shown by the zero at B's end of the sanction. Whereas B holds a single sanction, A can send one or more resources to B, and, as the signs show, each sanction is a loss of one to A and a gain of one to B. A's sanctions are like money. The number of sanctions that

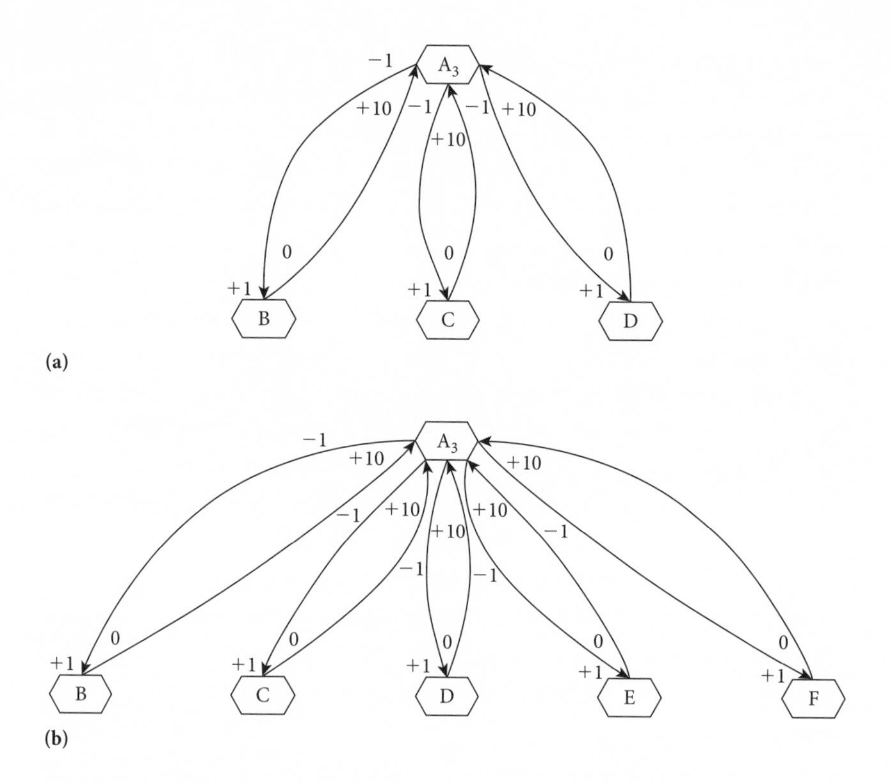

FIGURE 4.6. Null and Exclusively Connected Exchange Structures

A will pay to receive B's sanction is x. In fact, A has identical relations with the three peripherals and A's subscript of three indicates that A can exchange with all three peripherals.

Under what conditions will these actors transmit and receive sanctions and what payoffs will each gain? From the figure we can identify P_A, the payoff to A and P_B, the payoff to B. They are $P_A = 10 - x$ and $P_B = x$. That is, A's payoff is the 10 received from B's sanction minus the number of sanctions sent to B. B's payoff is simply the number of sanctions received from A. The same payoffs hold for A's relations with C and D. Elementary Theory applies the rationality principle discussed above. Applying that principle to the A–B relation, we can identify their negotiation set—the range of exchanges in which rational actors will engage. The negotiation set includes all exchanges for which both P_A and P_B are positive. Assume that A's sanctions are divisible into units of one. When A receives the sanction from B and sends one sanction in return $P_A = 10 - 1 = 9$ and $P_B = 1 - 0 = 1$. Because P_B receives only one, this is one end of the negotiation set. At the other extreme, A receives the sanction from B and sends 9 sanctions in return. $P_A = 10 - 9 = 1$ and $P_B = 9 - 0 = 9$. Sending 9 sanctions is the other end of the negotiation set. It also follows from the diagram that the same range of exchanges holds for the A–C and A–D relations. All exchanges occur within the negotiation set but where in that set? Something besides the rationality principle is needed to predict the point within that range at which exchange will occur.

Elementary Theory includes a resistance law and principle that predict which exchange will occur in the Figure 4.6a and 4.6b networks. $P_A max$ is A's best payoff while $P_A con$ is A's payoff at confrontation, when agreement is not reached. Resistance is

$$R_A = \frac{P_A max - P_A}{P_A - P_A con} \tag{4}$$

where the resistance factor to the right of equation 4 captures the mixed motives of an actor who (1) *competes* to gain higher payoffs as given by the factor in the numerator, but (2) *cooperates* to arrive at agreements as given by the factor in the denominator. Elementary Theory's second principle asserts that actors agree when their resistance factors are equal. For the A–B relation

$$R_A = \frac{P_A max - P_A}{P_A - P_A con} = \frac{P_B max - P_B}{P_B - P_B con} = R_B \tag{5}$$

A's best payoff, $P_A max = 9$ is at one end of the negotiation set and B's best payoff, $P_B max = 9$ is at the other. When A and B do not agree (confrontation), no sanctions flow and neither gains a payoff. Therefore, the payoff at confrontation, $Pcon = 0$ for A and B. Substituting these values into the resistance equation, $x = 5$, $P_A = 10 - 5 = 5$, and $P_B = 5$. That is, Elementary Theory predicts that A exchanges 5 sanctions for the one sanction from B. Because all three of A's relations are identical, it follows that the same prediction holds for A's exchanges with C and D.

$$R_A = \frac{9 - (10 - x)}{(10 - x) - 0} = \frac{9 - x}{x - 0} = R_B \tag{6}$$

The network in 4.6b is composed of the same kind of relations as 4.6a, but now the structural condition, *exclusion* affects exchanges and payoffs. A can still make three exchanges, but now has five partners among whom to choose. We describe the dynamics of this structure after A has reached (tentative) agreements with B, C, and D exactly like those calculated above. Because A has three offers, E and F face exclusion, not exchanging with A, and earning a zero payoff. Because A is rational, E and F must make better offers to A than did B, C, and D. The better offer is to ask for one resource less from A: then $x = 4$. Consider the A–E relation. If A fails to exchange with E, A will receive the 5 already agreed to with B, C, or D. Therefore, $P_A con = 5$, the payoff at confrontation, for the A–E relation. The best payoff that E can hope to gain is at $x = 4$: thus $P_E max = 4$.

$$R_A = \frac{9 - (10 - x)}{(10 - x) - 5} = \frac{4 - x}{x - 0} = R_E \tag{7}$$

Solving the resistance equation, $x = 2.5$. Therefore, $P_A = 7.5$ and $P_E = 2.5$. It also follows that F will make the same offer as did E and, with that offer, so will one of B, C, or D. Once A has received three $x = 2.5$ offers, only a better offer to A will avert exclusion for the peripheral. It follows that A will receive better and better offers with P_A increasing until $P_A = P_A max = 9$. Correspondingly, P values for peripherals decline to one. Thus, the three peripherals who exchange each gain one while the two who are excluded gain zero.

Brennan (1981) used the networks modeled in Figure 4.6 to design a pair of replicas to test whether exclusion has the effect predicted for the 4.6b network, and whether payoffs predicted for 4.6a, where exclusion is absent, will be observed. The models were plans for the replicas and thus for the physical layout of the experiment room. Seats for subjects were placed at each letter. Each exchange relation was built in the following way. Small tables were placed between the A chair and each of the peripheral chairs. Counters were stacked on each table. Rules for the experiment explained that B's stack of 10 blue counters was to be sent as a unit—and similarly for each of the other peripherals. The rules also indicated that the counters were worth nothing to any peripheral and one each to A. A had a stack of 33 counters. Each counter was worth one to A and one to the peripheral to whom it was sent. For both structures, all subjects were told that A could make, at most, three agreements. Only negotiations along the indicated relations are permitted and peripherals were separated by dividers to eliminate negotiations that might develop among them.

To generate preferences like those of the models, subjects were (1) paid by points gained in their exchanges, (2) asked to gain as many points for themselves as they could, and (3) told not to be overly concerned with the points gained by others.

Brennan conducted each experimental session in the following way. Each structure began with four rounds of negotiation and exchange. The rounds were timed by the experimenter who was present in the research room. Each round lasted 3 minutes and 20 seconds: subjects were told when 3 minutes had elapsed, which allowed them to complete pending exchanges. When the time had elapsed or when all exchanges that could occur had been completed—whichever came first—the experimenter redistributed the resources to their initial positions and began the next round. At the end of the fourth round, the first period ended. Then all subjects were rotated: A moved to the B chair, B to the C chair, and similarly with F moving to the middle to act as A. Then the second period began and was conducted exactly like the first. The second period was followed by a third and similarly until each subject had occupied each position for one period. Thus the experiment on the 4.6a network that had four subjects had four periods and the 4.6b network that had six subjects had six periods. Note that rotation preserves the structure while canceling effects of individual differences in negotiation skills, status cues, or other idiosyncratic differences.

For the 4.6a network, the mean payoff to the A from each exchange was 3.1 and 6.9 for peripherals. This result did not fit predictions particularly well. For the 4.6b network, A's mean payoff was 7.9 while each exchanging peripheral gained an average of 2.1. Examination of 4.6b network data showed that A's mean payoff fell short of $P_A = P_A max = 9$ because, although payoffs were increasing across rounds, they did not immediately reach that extreme. In fact, the modal payoff to A was the predicted $P_A = 9$. Thus, the results for the 4.6b network actually fit the prediction well. Remember that the aim of the experiment was to demonstrate the effect of exclusion and to that end it was a success. When exclusion was present A averaged 7.9 points and when it was absent averaged 3.1, thus gaining more than twice as much when exclusion was present than when it was not.

On the other hand, for the 4.6a network, the A–B exchange of 3.1–6.9 diverged far from the predicted 5–5 payoffs. Either the theory under test is unsound or the experiments did not exactly replicate the models. If the theory is right, the observed result could have been produced if subjects introduced sanctions that were not included in the model. For example, peripheral subjects could have negatively sanctioned A with complaints and insults. Perhaps subjects in peripheral positions contingently sanctioned A in the hope of inducing better offers. If so, they could have pushed the observed exchanges away from the predicted value. On further inspection, it was apparent that peripherals had introduced sanctions not in the model. When they got low payoffs, they complained to the central subject that she was greedy.

Is there a material foundation on which the subjects could claim that the central subject was greedy? If exchanges occur at the predicted exchange ratio, both central and peripheral subjects gain 5 points from each exchange. Nevertheless, at the end of each round, peripherals each had only the 5 points, the value of the 5 sanctions sent by the central subject. By contrast, the central subject had $10 + 10 + 10 = 30$ points from the 3 sanctions sent by peripherals and also had $33 - 5 - 5 - 5 = 18$ additional points, which was the value of the resources left over from the 33 initially held. Therefore, at the predicted exchange ratio, the central subjects had $30 + 18 = 48$ points after each round (an average of 16 points per exchange) while each peripheral had only 5 points.[9]

A revised scoring system was created to eliminate the basis for the peripherals' introduction of sanctions not in the model. The new system reversed all exchange relations in the following way. A held the initially worthless sanctions and the Bs held the resources, which, like money, were valuable if retained. Now, at the predicted $x = 5$ exchange, the central A held $5 + 5 + 5 = 15$ points at the end of each round. Each peripheral also held 15 points: 10 points from the sanctions A sent plus the 5 not spent in exchange. This design (1) dampened out the complaints from peripherals and (2) resulted in mean exchanges not significantly different from $x = 5$ predicted by theory. While this modification supports predictions, it does not rule out the possibility that equity norms also had a role in producing the $x = 5$ result.

Now we apply the five maxims to Brennan's experiments. Again the maxims are given in italics:

1. *Derive one or more models from the theory to be tested.* Two models were derived from the theory and its laws. The first model was a structure without exclusion while the second was with exclusion.
2. *Use the theory to generate predictions by linking initial conditions to end conditions.* The theory's resistance law and resistance principle coupled with the rationality principle offer contrasting predictions for the two models. They predict medial exchange ratios in structures without exclusion, but extreme ratios when exclusion is present.
3. *Build replicas, set initial conditions, and observe the end conditions.* In the layout of the experimental room, the experimental replica

looked like the model for it. In addition, subject payments and experimental demand were designed to bring subjects' preferences in line with conditions of the models. Subjects faced each other over stacks of resources that could be exchanged on joint agreement. The two designs differed on only one initial condition, the number of peripherals (either 3 or 5). Since A was permitted only 3 exchanges per round in each structure, the structure with 5 peripherals has 2 exclusions.

4. *Compare results to predictions and decide whether the theory is supported.* Results corresponded to predictions for the 4.6b structure with exclusion. Predictions for structure 4.6a were not supported but reapplying the theory and creating new initial conditions led to new experiments for which predictions were supported.
5. *Make inferences from theory with greatest confidence to instances most theoretically similar to experiments supporting the theory—but predictions are not formally limited by that similarity.*

Regarding the fifth point, the experiments described here are the first of many studies that have been based on applications of Elementary Theory. Exclusion occurs when $N > M$, where N is the number of relations with which a position can exchange and M is the maximum number of exchanges that it can complete. For the 4.6b structure $N_A = 5 > 3 = M_A$. Exclusion is only one of seven structural conditions of power discovered by elementary theorists and experimentally investigated by them. A *null* connection is one for which $N = M > Q = 1$, where Q is the number of exchanges a position must complete before it can benefit from any exchange. Because $N_A = 3 = 3 = M_A > Q_A = 1$, the 4.6a network is null connected. (Q is also equal to one for the exclusive connection of the 4.6b structure.) Because descriptions of all seven structural conditions and experiments investigating them are available elsewhere (Walker et al. 2000; Willer 1999; Corra and Willer 2002), they will not be discussed further here.

As the scope of Elementary Theory has been extended, its core concepts have remained few. For example, predictions for all seven power conditions are generated using the rationality principle, the resistance law, and the resistance principle—and nothing more. Those same theoretic ideas are used to make predictions for coercive structures (Willer 1987) and for exchange structures in which resources flow in multiple steps through a network (Willer 2003). These extensions suggest that the study of social structures has only begun and that there is far more to be learned in the future. Fortunately, the software to build experiments is available without cost on the Web (http://weblab.ship.edu). It could be used, for example, to replicate Brennan's exchange experiments. That software and some of its uses will be discussed in Chapter 6.

Summing Up

This chapter began with the assertion that theory-driven experiments in sociology are methodologically identical to experiments in other sciences. They are identical in that theory designs the experiment. Two proofs have been offered. First, the chapter has presented three pairs of experiments wherein each pair is composed of one from physics and one from sociology. The empirically inclined among us will assuredly find that the paired experiments are more similar in spite of their contrasting fields than are those from the same field. There is one exception. The second experiment used Elementary Theory as did the last discussed. Thus those two are similar to each other.

For the theoretically inclined among us, a second, stronger proof was offered. The principles of theory-driven experimentation were set forth early in the chapter and applied as maxims to all six experiments discussed. The maxims fit physics and sociology experiments equally well. The idea fundamental to both is to build experimental replicas to be like models drawn from the theory. Then the experiment is a test of the similarity of the two from initial through end conditions.

Because theory-driven experiments in sociology are not unique from or inferior to those of other sciences, it is high time to put aside the notion that the utility of experiments is limited because their results cannot be generalized. No one should ever generalize from a theory-driven *experiment*. Instead, the correct scientific procedure is to apply experimentally tested theory to cases outside the laboratory.

For example, Elizabeth G. Cohen, a pioneering SCT researcher, applied SCT outside the laboratory. Cohen recognized that reading ability is a status characteristic (Rosenholtz and Cohen 1983). She also realized that students differ in reading ability. Using SCT, Cohen predicted that other students *and* teachers would use reading ability (just as they use other characteristics, like race or gender) to develop expectations about students' intellectual skills. Moreover, they would use characteristics like race or gender to develop expectations about reading skills. In turn, students and teachers interact with students based on their actual or attributed reading skills in a way that tends to verify self-fulfilling prophecies about how well students would do at school. Thus, low status children come to read less well than high status children. Cohen did not stop there; she used SCT to devise strategies to intervene in classrooms, to disrupt those processes, and to help low status students learn better (Cohen 1998; Cohen, Lotan, Scarloss, and Arellano 1999).

Cohen brought the research process on status and influence full circle. It began with questions, devised answers to those questions in the form of theory, used experiments like Moore's to test the theories, and finally showed that experimentally supported theory can solve problems outside the lab. Now, we will turn, in Chapter 5, to the practical issues involved in building experiments. In Chapter 6 more will be said about linking experimental and nonexperimental research through theory, including applications of Elementary Theory outside the lab.

5 EXPERIMENTS AS SOCIAL RELATIONS

IN THE SOCIAL AND BEHAVIORAL SCIENCES, EXPERIMENTAL DEsigns create social situations that contain social relationships. Within those relations as Martin T. Orne (1962:777) put it, "The roles of subject and experimenter are well understood and carry with them well-defined mutual role expectations." The experimenter-subject relationship is governed by culturally and institutionally defined rules and norms that guide and constrain the actions of the role occupants. It follows that the experimenter-subject relationship affects every stage of the research process and social scientists must consider how the *social* character of the relationship affects the processes under study.

In the experimenter-subject relationship, experimenters typically have higher status, greater power, and more authority than subjects. That inequality has important consequences for subjects and for the experiments in which they are studied. Experimenters can misuse their greater power and authority to exploit subjects. They can influence subjects to do things they would be unwilling to do outside the research setting. We take up that issue in the first section of this chapter, which discusses ethical standards—norms of science that define the rights of human subjects and the responsibilities of researchers.

Subject recruitment establishes the experimenter-subject relationship. Subject recruitment is worthy of study especially because the time and effort it requires is of a similar magnitude to the time and effort expended in actually running an experiment. We examine two recruitment issues, beginning with the nuts and bolts of how to recruit subjects. We then turn to ethical issues focusing on subject payments.

The experimenter's greater influence and power in the experimental setting can also affect the outcomes of experimental research in unintended and unnoticed ways. These effects, which may be either consistent or inconsistent with hypotheses or theories that guide the research, are discussed as artifacts, demand characteristics, and experimenter bias. That discussion explores the ways that research settings, experimenters' and subjects' reactions to settings, and experimenter biases can affect research outcomes.

The chapter concludes with an introduction to modern computer-mediated experimental settings. These settings are intended to optimize control over the social relations of sociological experiments and frequently do so. They also improve the fit between theoretic models and experimental replicas while allowing replications to be as identical as possible. Both experimenter and subject biases are reduced by interposing a computer interface between experimenter and subject. In fact, improving fit, creating exact replications, and mediating experimenter-subject interaction should reduce all of the kinds of artifacts including demand characteristics and experimenter bias that are discussed in this chapter. Fortunately, the software for both examples we discuss is available without fee to interested scholars.

First Do No Harm: Ethics and Research with Human Subjects

Perhaps only comedians have expressed concern for the physical or emotional well-being of the subatomic particles that physicists smash against particle screens. By contrast, medical, social, and behavioral researchers can do lasting physical, psychological, and emotional harm to the people they study. Because they can, they have a responsibility to ensure that they do not compromise the well-being of human subjects. The medical profession's dictum: "First, do no harm" is a good rule for all who deal with human subjects.[1] As we now show, in social and medical science many have failed to follow that rule.

Two Studies

The Tuskegee Syphilis Study The National Health Service began a study in Macon County, Georgia, in 1932. The Tuskegee Syphilis Study involved 600 black men, 399 of whom had syphilis when the study began and 201 who did not. There was no reliable cure for the disease, and poor, undereducated people were most affected by it. The *best* treatments available at the time used highly toxic drugs and the cure rate hovered around 30 percent. The National Health Service promised study participants that their health status would be monitored while they were being treated for the disease.

The Tuskegee Syphilis Study was actually a method-of-difference experiment designed to observe the long-term health effects of *untreated* syphilis. Researchers compared the health status of infected participants with those who were not infected. Penicillin became the standard treatment for the disease in the late 1940s but men in the Tuskegee Study were never given that drug. Moreover, neither the men nor the general public were aware of the true nature of the study until the *New York Times* broke the story in July 1972—40 years after the study began (Jones 1981).[2]

The Stanford Prison Experiment In mid-August of 1971, a year before details of the Tuskegee study were revealed to the nation, police officers in Palo Alto, California, arrested and jailed nine summer school students. The students, all male, were "arrested" as part of the Stanford Prison Experiment (SPE) directed by psychologist Philip Zimbardo (Haney, Banks, and Zimbardo 1973).[3] By Zimbardo's account, he and his fellow researchers wondered "what would happen if we aggregated all these processes making some participants feel deindividuated and others dehumanized within an anonymous environment that constituted a 'total environment' in a controlled experimental setting" (Zimbardo, Maslach, and Haney 2000:9).

Zimbardo and his colleagues screened 70 student volunteers and pared the numbers to two dozen men who seemed normal. Half the men were chosen at random to be prisoners. The other half became guards in a makeshift prison created in the basement of the psychology department at Stanford University. Zimbardo and his colleagues conducted their study in a *controlled environment*. The researchers did not control features of the situation like deindividuation or dehumanization (Zimbardo et al. 2000:9), and we are unable to identify any theory or theoretical model under test. Consequently, it is neither a method-of-difference nor a theory-driven experiment.

The researchers stopped the SPE after only 6 days although it was planned to continue for up to 2 weeks. During that 6-day period, prisoners rebelled and guards became hostile and brutal. At times, the guards broke the rules they had established for prisoner treatment (e.g., keeping prisoners in solitary confinement for longer periods than their rules allowed). Some parents intervened and sought legal counsel to get their sons out of "prison." Moreover, Zimbardo and other "role players" behaved less like researchers in search of understanding and more like heavy-handed prison authorities quelling unrest among inmates.

The Tuskegee study and the Stanford Prison Experiment violate the principle of *primum non nocere.* Many Tuskegee volunteers suffered needlessly with syphilis after a cure was available, and many died premature and painful deaths because they were not treated. Some of the men who were in the SPE had severe psychological and emotional reactions, and five were released before the researchers shut the study down. These studies and countless others like them raise troubling ethical questions. What are the responsibilities of researchers to their human subjects? What rights do human research subjects have?

The Rights of Human Subjects

Every investigator who conducts research involving human subjects is responsible for their protection. The scientific community has established what amounts to a "Bill of Rights" for human participants in medical and behavioral research. The National Institutes of Health's Belmont Report (National Institutes of Health 1979) established a broad framework that outlines the rights of human subjects and researchers' responsibilities to them. Their rights include but are not limited to the following:

1. *Subjects have the right to be protected from unwarranted risk.* Risks include potential emotional, psychological, or physical harm.
2. *Subjects have the right to be fully informed of research procedures.*
3. *Subjects have the right to be fully informed of any and all potential risks.* Researchers must protect subjects from harm, and they are also required to tell subjects about any known risks of their research procedures. Telling about known risks, no matter how small, allows subjects to make informed decisions about beginning or continuing participation in a study.
4. *Subjects have the right to refuse to participate in a research project.* Some academic researchers ask students to participate in one or more research projects in exchange for partial course credit. However, students—like all other human subjects—have an absolute right to refuse to participate in a research project.
5. *Subjects have the right to withdraw from a research project at any time.* A research subject can agree to participate in a study and decide

to end her participation before the study is completed. Under such circumstances she cannot be compelled to continue.

6. *Subjects have the right to be protected from coercive participation.* Subjects cannot be coerced either psychologically or physically to participate involuntarily in a study.

Ethics Enforcement Federal regulations and local practice at research organizations require investigators to (1) be certified to do research with human subjects and (2) have an approved plan for protecting subjects in their investigation. As part of the certification process, researchers are required to demonstrate their familiarity with the rights of subjects, researchers' responsibilities to them, and with procedures used to protect subjects' rights. Many institutions, including our home universities, require researchers to pass examinations that cover material developed at the University of Rochester (Dunn and Chadwick 2001, sometimes called the Rochester protocol).

Concern for human subjects is not limited to federal oversight of research activities. Professional organizations of the human sciences have extensive codes of ethics that govern the actions of practitioners in their respective fields. We will refer to several sections of the American Sociological Association's Code of Ethics (1999) in the discussion that follows.[4]

Designing and Conducting Ethical Research

Researchers planning studies with human subjects should contact their local institutional review board (IRB) and ask about its procedures for protecting human subjects. Most IRBs are organizational units of institutions that are routinely engaged in research with human subjects (e.g., research universities or medical centers). Some IRBs are independent, freestanding organizations that review projects for independent research contractors or for research organizations that are too small to have their own review boards. Researchers, including research assistants, should begin training programs or test preparation to secure certification well in advance of the proposed study's start date. Normally, researchers who apply for government or foundation support are required to provide evidence of certification before any monies can be released to their project.

Every research project must devise a plan for protecting the rights of human subjects and submit it to an IRB.[5] Research plans submitted to an IRB must describe in detail how the proposed research ensures protection of participants' rights. Plans must also show how researchers will verify that participants are informed of their rights *before* they begin participation. IRB approval must be secured *before* conducting research. Investigators are also required to inform the IRB or one of its officers of complaints made against the project. There are stiff penalties for failure to comply with these ethical guidelines, including fines, loss of research funds, and ineligibility to apply for research funds.

Ethical Standards in Practice: Deception in Social Research

Men in the Tuskegee experiment did not know that their disease was not being treated or that after 1947 there was an effective treatment for syphilis. Subjects in Moore's experiment (Chapter 4) did not know that they were "interacting" with a machine. These studies seem to violate several provisions of the subjects' Bill of Rights. First, subjects in these studies were not given complete and accurate information about research procedures. As a result, their consent to participate could not be fully informed. Second, men in the Tuskegee study were not told of the potential risks of participation. Furthermore, Tuskegee and Stanford Prison researchers failed to protect their subjects from physical and emotional harm. Men in both studies were harmed either temporarily or permanently and the risks could have (or should have) been foreseen. In contrast, there is no evidence that subjects in the Moore study were harmed in any way.

Is deception like that practiced by Moore legitimate? Many researchers use deception and misdirection in their experiments, but others are unalterably opposed to the practice (Geller 1982; Kelman 1967). Deception raises difficult ethical concerns. It appears to violate the requirement of full information (but see below), and it can conceal potential risks from subjects. Still, neither Asch nor James Moore began their studies by telling subjects that they would get false feedback from experimenters or their confederates. Telling subjects that there would be false feedback would defeat the purpose of these studies. It is unlikely that their subjects would have taken the studies seriously had they done so. How can researchers comply with ethical standards in disclosing research procedures—including deception—and protect the integrity of their experiments?

The codes of conduct of the major social science professional associations (see endnote 4) regulate when deception is and is not ethical. The standards established by the American Sociological Association (ASA) permit deception in research if three conditions are satisfied. Researchers can use deception if:

> [it is] determined that [deception] will (1) not be harmful to research participants; (2) [deception] is justified by the study's prospective scientific, educational, or applied value, and (3) that *equally effective alternative procedures that do not use deception are not feasible.* (ASA 1999:14, section 12.05a, emphasis added)

Researchers whose studies satisfy these criteria use deception and most tell subjects about deceptions in post-session debriefing—after critical parts of the study have been completed. Moore followed the practice now mandated by ASA policy, which declares that deceptions must be revealed *no later than the end of a study* (ASA 1999:14, section 12.05c). Still researchers must weigh carefully the potential risks and benefits of deceptive research.

Types of Deception Psychologists and other social and behavioral scientists have studied reactions to deception. Their research shows that both deception and the post-session disclosure of deception can put *subjects and research staff* at risk of emotional or psychological damage. Research on deception also includes evaluations of techniques used to reverse and eliminate the potentially harmful effects of deception. Sieber (cited in Geller 1982; see also Sieber 1992) identifies three forms of deception used in research. They are *implicit*, *technical*, and *role* deception.

Implicit deception occurs when subjects do not know that they are in a study. The Tuskegee Syphilis Study is a notorious example of implicit deception (Jones 1981). Social science has also produced bad examples of implicit deception. Laud Humphreys (1970) acted as a lookout while men engaged in homosexual encounters in public restrooms. He also copied their license numbers and got their addresses from the Department of Motor Vehicles. None of the men knew that Humphreys was a researcher until he appeared at their residences and asked for interviews about their personal and family lives a year or so after he had observed their homosexual encounters.[6]

In other cases of implicit deception, it is likely that some "participants" went to their graves without knowing that they had taken part in a study. For example, Piliavin, Rodin, and Piliavin (1969) observed bystanders' reactions to confederates who feigned collapse on New York subway trains. The participants in those experiments could only learn of the study if they read articles in scientific journals or happened to read popular reports of *bystander intervention* research. Research in public places (like that of Piliavin et al.) does not require consent if the case can be made that there is no or minimal risk to subjects (ASA 1999:12, section 12.01c).

Researchers who disguise the role of equipment or procedures engage in technical deception. Moore told subjects in the SCT experiment described in Chapter 4 that lights on the Interaction Control Machine (ICOM) recorded their own and their partner's initial choices. In fact, the experimenter controlled the ICOM to ensure that 70 percent of the paired choices disagreed. Milgram's claim that his shock generator delivered an electrical shock to a learner is another example of technical deception in laboratory studies.

Role deception occurs when researchers misidentify the roles of others in a study. Seven of the "subjects" in Asch's original studies were confederates whose role was to pressure subjects to give incorrect answers. The "learner" in Milgram's obedience studies was always a confederate.

Harmful Effects of Deception Deception can do psychological damage to subjects and researchers. Imagine that a researcher has just told you that you are the victim of a scam. The partner to whom you have been sending messages for the last 45 minutes does not exist. Alternatively, you have just been told that the other subjects are not subjects at all, but working for the experimenter and their only reason for being there is to mislead you. Would you feel dumb? Embarrassed? Humiliated? Angry? Anyone might experience any or all of these feelings after having learned that she has been deceived. The subject has been taken in by the experimenter and, as a result, there can be deep and lasting effects.

Some deceptions get subjects to take actions that are, on their face, cruel, unusual, or inconsistent with the subjects' personal values. Milgram's obedience experiments are a classic example of such research as is that of Zimbardo, Milgram's high school classmate (Zimbardo et al. 2000). Most of us have done—or failed to do—some things that we are not proud of. Sometimes we have second thoughts about such events. Should I have dialed 911 when I heard that scream in the middle of the night? What kind of person could I be if I failed to stop at the scene of a serious auto accident? In many cases, we justify our behavior by telling ourselves that we "had no time to think" because it happened so quickly. In other cases, we assume that someone else would do something about the problem. Psychologists call the latter response *diffusion of responsibility.*

The potentially harmful effects of deception are not limited to subjects who learn that researchers have deceived them. Research assistants can have similar reactions. After all, what kind of person continues to do a job that involves lying to people every day or, worse, watching normal college students descend to the level of animals in a make-believe prison?[7]

Professional ethics are clear and most researchers agree that experimenters should, whenever possible, avoid creating and using research designs that deceive subjects (Kelman 1967; Geller 1982). Still, some investigations may not be possible without deception. For example, medical researchers give some subjects placebos in a series of clinical trials. Without placebos, researchers might be unable to determine if changes in health status are bona fide medical effects or social-psychological effects of research participation. However, researchers who use misdirection and deception have a moral and ethical responsibility to ensure that negative reactions to deception, if any, are only temporary. It is their duty and ethical responsibility to ensure that subjects and research assistants are not emotionally damaged by their research experiences. Debriefing helps researchers accomplish that objective.

Debriefing

> It is believed by many undergraduates that psychologists are intentionally deceptive in most experiments. If undergraduates believe the same about economists, we have lost control. It is for this reason that modern experimental economists have been carefully nurturing a reputation for absolute honesty in all their experiments. This may require costlier experiments. (Ledyard 1995:134)

While we do not disagree with the spirit of Ledyard's comments, he misunderstands when deception can be and is used. The ethical standards are clear. At issue is not the cost of the experiment, but whether it can be done without deception. Only in the latter case can deception be used. Furthermore, the use of deception does not necessarily lead to loss of experimental control.[8] A good debriefing brings the subject into the experimental process as a partner. As a part of that partnership, subjects may be asked to hold the deception in confidence and very frequently they will do so.

Post-experimental debriefing has two main objectives. First, it ensures that researchers fully inform subjects about the research in which they have participated. Second, it gives researchers the opportunity to restore the participants to the same psychological and emotional state they were in when they entered the study.

Post-study debriefing should be an integral component of all deception research with human subjects. Researchers must *dehoax* and *desensitize* subjects who have been deceived (Holmes 1976a, 1976b). Dehoaxing focuses on the experimenter's behavior (e.g., on deceptive procedures). Desensitization centers on the subject's behavior including new or altered cognitions that are created or encouraged by deceptions or by the experimenter's disclosure of deceptive practices. Neither task is simple. We use the research of Dov Cohen and his colleagues as an example.

Cohen, Nisbett, Bowdle, and Schwarz (1996) designed a series of method-of-difference experiments (1) to learn if there is a southern "culture of honor" and (2) to find whether the culture of honor is related to male aggressiveness. Cohen et al. studied male students at the University of Michigan in a 2 × 2 experimental design with subjects' region of origin (South vs. non-South) and insult (insulted vs. not insulted) as the experimental variables of interest. The researchers assumed that southern males would react more aggressively to insults than men from other regions *if there was a southern cultural of honor.*

Subjects in the experiment met with experimental assistants who gave them a brief description of the experiment and had them complete a short demographic questionnaire. After completing the questionnaire the subject was told to place it on a table at the end of a long narrow hallway. As the subject walked down the hallway, a male confederate who was standing at an open file drawer, slammed the file drawer closed, turned, and walked toward the subject. The confederate bumped the subject's shoulder as they passed and called the participant an "asshole." Two other confederates (one female and one male) who appeared to be studying witnessed the incident and recorded the subject's verbal and nonverbal reactions including his facial expression. After the incident, each subject completed several tasks including writing the endings of two scenarios. Subjects were dehoaxed and desensitized in the post-session debriefing.

Dehoaxing Researchers should begin debriefing sessions by dehoaxing subjects. In dehoaxing, researchers tell subjects about the deceptions they used and why they used them. Dehoaxing cannot be successful unless the research staff can convince subjects of their honesty during the debriefing process. After all, they have just revealed their dishonesty.

Cohen et al. began their debriefing by discussing the research design with the subjects. The subjects were given an opportunity to discuss the true purposes of the experiment after the researchers described the various deceptions (e.g., the roles of the three confederates) and why deceptions were important to the success or failure of the experiment. At that point, subjects knew that they had been deceived, how they had been deceived, and how the researcher expected them to react to the deception. The hoax was fully disclosed. Cohen et al.'s debriefing also included bringing the insulting confederate in and having him talk with the subjects.

Desensitization Desensitization is the next step. Deceptive research practices or the investigator's disclosure of deception can cause subjects to reconsider and change their self-definitions. If not corrected, altered self-concepts can influence study participants after they leave the lab. It is the debriefer's job to restore the subjects to their pre-experimental condition. Cohen and his associates had available to them the results of the post-session questionnaires they gave to subjects. They were aware of which subjects were still angry after the initial debriefing and dehoaxing.[9] Experienced investigators continue post-session debriefing until all the subject's questions are answered to the *subject's* satisfaction. Ethical investigators also tell subjects who remain upset after dehoaxing and desensitization where they can file complaints and to whom they can talk about psychic or emotional discomfort they feel after they leave the lab.

Debriefing procedures are established *before* a researcher puts any subject into an experimental situation. An IRB must approve all procedures before a researcher is permitted to bring any human subject into an experiment. As our discussion shows, researchers devote much thought, time, and energy into building designs that ensure the ethical treatment of subjects and, if deceptions are used, developing and using debriefing procedures that include dehoaxing and desensitization.

Ethical reasons alone should be enough to motivate researchers to avoid creating risks for those they study. Fortunately, ethical treatment is also called for on pragmatic grounds. We know of no research that shows that unethical studies produce scientific findings that are as useful as findings from research that adheres to strict ethical standards. Unethical studies like the Tuskegee study or the Stanford Prison Experiment are examples of bad science.

Recruiting Subjects

Where do subjects come from? How can a researcher gain access to them? It should be obvious that subjects will not arrive by accident at research labs. Some kind of inducement will have to be offered. And yet, one kind of inducement—demanding that students participate in an experiment as a course requirement—though long used is no longer available. Today, that demand is seen as coercive and thus unethical: most IRBs do not allow it. On the other hand, offering participation in experiments as one alternative for gaining extra course credit is ethical. Another way to induce prospective subjects to our labs is to pay them. Though paying subjects is a benefit to them, the practice can have pitfalls. We trace some of the pitfalls of paying below.

The time and effort expended on subject recruitment and scheduling is comparable to that expended in running experiments. Using our own work as an example, in a good week, we ran 72 subjects in 12 sessions. Total running time for the week was about 30 hours. Total time spent in recruiting and scheduling was just over 18 hours. We now trace some procedures we have used to recruit and schedule university student subjects.

While student newspaper ads can be used, we prefer to recruit in large classes. Fortunately, professors across many fields have proved to be very helpful in allowing us to come to their classrooms and sign up their students. Here the first step is to find undergraduate courses that enroll large numbers of students and contact those who teach them. We have found that instructors are most amenable to our recruiting at the start of their classes and we make it a point to always arrive early. Typically we make a brief statement prior to handing out forms explaining the study and whether there will be payments or course credit for participation. Our contact forms ask for name, telephone number, and email address as well as information on any previous experience in experiments.

With a list of prospective subjects in hand, the next task is scheduling. We use schedulers to arrange participation. Having been given a stack of contact forms, the scheduler phones or emails prospective subjects and arranges for the required number of subjects to come to the lab at a given time. Typically, schedulers are responsible for giving clear directions to the lab to ensure that everyone appears at the correct place at the correct time. Because it is very unusual for all scheduled students to actually appear at the lab, experiments that require multiple subjects must schedule more than the needed number. Because scheduling can occur as much as a week in advance of the study, we place phone calls to remind prospective subjects the night before the study.

In our departments, experimentalists work together to build a common subject pool from which all draw. Importantly, subjects who have participated in experiments that do not use deception can be used again on another experimental project. Because recruitment is such a time-consuming task, multiple scheduling of subjects is highly desirable.

Researchers in sociology and psychology only rarely recruit subjects from the general population. Milgram (1974) was an exception. He used newspaper ads to recruit adult volunteers from the communities around Yale. In the future, recruitment may fundamentally change. Internet-based experiments with open recruitment will undoubtedly become more and more common. These studies, ongoing on the net, are joined by anyone interested in participating and, as a result, offer the possibility of large data harvests at little or no cost. Only the future will tell us whether experimental control can be maintained and whether subjects' rights can be protected with open recruitment on the Internet.

We now turn to the issue of subject pay. Experiments analyzed earlier (those run by Moore, Brennan, and Simpson) paid subjects. Medical researchers, and more particularly, students of bioethics, have written extensively about the pros and cons of paying subjects. On the con side, some label payment of subjects in medical experimentation unethical because paid subjects are more likely to expose themselves to risks that they would otherwise avoid (McNeill 1997). Some claim that paying subjects invalidates the principle of informed consent because paid subjects may well ignore the information they have been given about risks. Both problems are greater if the potential subject is poor (McNeill 1993). And what if subjects are not paid unless they complete a study? Are payments then ethical?

On the pro side, Wilkinson and Moore (1997) offer a spirited defense of payment, but we will not detail their arguments here. Instead, we point out that for most—but perhaps not all—social science research, the amounts paid to subjects are not large enough to overcome information about risk or to invalidate informed consent. Moreover, the risks to subjects in most social science research are minimal. This is particularly true today because of the sensitivity of local IRBs to the protection of human subjects. Thus paying social science subjects is ethical, but withholding payments from subjects who complete only part of a study is not.

Artifacts: Demand Characteristics and Experimenter Bias

Experiments are designed to produce effects. Empiricist designs use Mill's method of difference and, if successful, find regularities between effects and one or more experimentally controlled variables. The objective of theory-driven experiments is to produce replicas of the structures and processes modeled by a theory (Willer 1987). However, some effects are artifacts of the experiment. Kruglanski (1975:103) describes an *artifact* as "an error of inference regarding the cause of an observed effect." More accurately, artifacts are systematic effects produced by (1) variables over which the investigator has not exercised control or (2) variables under the experimenter's control that create effects different than those the researcher or theory anticipates.

Artifacts can lead researchers to make errors of inference because they affect experimental results. In some cases, the effects of artifacts are indistinguishable from the effects of variables that measure theoretical constructs. Demand characteristics (Orne 1962) and experimenter bias are important sources of artifacts. We define both as they are taken up below. We now show that researchers can minimize demand characteristics and experimenter bias in experimental designs in three ways. First, they can give careful attention to these issues at the design phase. Second, they can build methods for detecting sources of artifacts into experimental protocols. Third, they can systematically pretest new designs.

Demand Characteristics

Orne coined the term "demand characteristics" to describe all the cues in an experimental situation that telegraph the study's hypotheses (Orne 1962:779, 1969). Orne pointed out that the experimenter-subject relationship could lead subjects to have expectations about how a "good subject" ought to behave even if they were unsure about the specifics of good subject behavior. Good subjects have a stake in the experiment's success. They are concerned about whether the study's hypotheses *turn out right*. The result is *subject bias*. A subject who reacts to demand characteristics either intentionally or unintentionally biases the study's outcomes.

The problem is more complicated than Orne's discussion implies. There are at least four possibilities, and we treat all of them as demand characteristics.

1. Subjects correctly infer experimental hypotheses and try to produce expected results. This is the case Orne describes.
2. Subjects correctly infer experimental hypotheses and try to produce contrary effects. Weber and Cook (1972) raised this possibility.
3. Subjects *incorrectly* infer experimental hypotheses and try to produce the effects they infer.
4. Subjects incorrectly infer experimental hypotheses and try to produce contrary effects.

It follows from the four possibilities that demand characteristics can produce effects that are either consistent or inconsistent with predicted effects. Whatever their direction, demand characteristics lead to errors of interpretation. Suppose that demand characteristics produce effects consistent with theoretical predictions. Whether the experimental conditions could have produced the effect or not, the conclusion that the study supports theoretical predictions is erroneous.

Demand characteristics that produce *inconsistent* effects can also lead to errors of interpretation. Consider situations in which variables under experimenter control produce effects consistent with theoretical predictions, but demand characteristics produce equally strong but inconsistent counterbalancing effects. This outcome can lead a researcher to conclude that theoretical variables had no effect and that the results do not support her theory. In another case, demand characteristics might produce strong effects that overwhelm the effects controlled by the experimental treatments. Now the researcher who is unaware will conclude that the theoretical variables have an effect that is exactly opposite to that predicted by the theory.

According to Orne, demand characteristics cannot be eliminated because experimenters cannot control *all* of the interests subjects bring to an experiment. Willer (1987:191) identified three conditions that are necessary for demand characteristics to produce artifacts and we modify them to take into account the four possibilities just given: (1) The subject must infer a hypothesis; (2) the subject must want to produce a result consistent (or inconsistent) with the hypothesis; (3) the subject must be able to produce the result. Orne concentrated on the first condition and claimed that detection through post-session debriefing was the optimal technique for detecting demand characteristics. Detection is important but researchers can also take steps to reduce and manage demand characteristics. We return to Cohen et al.'s (1996) study to show how their design controls subject bias by eliminating the bias opportunity.

Consistent with the broader notion of artifacts we are using (i.e., effects of uncontrolled variables), Cohen et al.'s design included several features that eliminate potential sources of artifacts. First, the researchers assumed that people who "left home" to attend school might be different in many ways than those who remained in their home state. Therefore, University of Michigan students who were Michigan residents were excluded from the experiment. Second, Jewish students were also excluded under the assumption that Jewish cultural traditions could be another source of influence on the behavior of southern and non-southern men.[10] Third, the insult was always the same and it was always delivered after the confederate had physical contact (the bump) with the subject. Fourth, and *here is where the bias opportunity is eliminated*, the subjects were not aware that the insult, the variable under experimental control, was part of the experiment. Because they were not, no attitude toward the experiment could influence its result and demand characteristics could have no effect.

Other designs reduce the possibility of artifacts based on demand characteristics. When a subject is participating alone in an experiment, it may not be possible to distinguish behavior that is a consequence of the independent variable (or initial conditions) from behavior stemming from demand characteristics that is motivated to prove or disprove real or imagined hypotheses. However, consider any sociological experiment in which subjects interact as they do in the exchange network studies described in earlier chapters. Assume that *all* subjects in those experimental groups had inferred the predicted result, the exchange ratios predicted by theory. Assume further that all took on the role of good subjects to produce the result by *role-playing alone*. They could produce the predicted exchange ratios, but they could not produce the complex interaction process leading up to them. Exchange ratios in network exchange studies are produced through an interaction process well known to experimentalists. Only two scenarios are possible for role-playing subjects. First, the interaction process would be wholly absent. Subjects would set the predicted rates without negotiation. Alternatively, the interaction process would proceed as if the subjects were in a very bad play.

Interactions that follow from and conform to processes described by network exchange theory will be very different than interactions of subjects engaged in role-playing to confirm hypotheses. Role-playing interactions cannot be rehearsed and the average subject is not an accomplished actor. The result will not look like negotiations. Anyone who doubts that a role-played interaction would be very different should attend an early rehearsal (or audition) of a play and evaluate the realism of the actors in their roles.[11]

Experimenter Bias

Demand characteristics are real or imagined features of experimental situations that produce subject bias. Experimenters can also influence subject behavior in ways they do not intend. Frequently, researchers enter the laboratory with full knowledge of the research hypotheses and, in many instances, so do their research assistants. An experimenter who knows what the hypotheses are may talk, stand, or walk in such a way as to give off information about those hypotheses. When verbal and nonverbal cues of the experimenter, whether subtle or not, affect a subject's behavior, the experiment suffers from experimenter bias. Experimenter bias, like demand characteristics, can make interpretation of research findings difficult or impossible.

Robert Rosenthal (1966) used an experiment to demonstrate the power of experimenter bias. Three groups of students assisted in a study that involved laboratory rats running mazes. One group was told that their rats had average abilities. Groups two and three were told that their rats were smart and dumb respectively. Students who ran dumb rats consistently reported poor performances whereas those who had smart rats reported high performances. Outside the lab, bias of this sort can also have important practical consequences. Rosenthal and Jacobson (1968) studied teacher expectations and found that teachers who expected poor performances from their students got poor performances. Those who expected good performances had higher-performing students.

Milgram's obedience experiments were exercises in experimenter bias. Milgram seemed to be working without hypotheses—he certainly had no theory—but his active involvement with subjects appeared to be related to the high levels of obedience.[12] Obedience was negligible when (1) ordinary people gave the subjects orders or (2) subjects could choose the level of shock.

The possibility of experimenter bias is reduced substantially by running "blinded experiments." An experimenter is "blind" if she does not know the experimental treatment under study. For example, neither the insulting confederate nor the confederate observers were aware of the subject's region of origin in the Cohen et al. (1996) studies described above.

A variant of blind experiments, the double-blind experiment, ensures that neither the experimenter nor the subject is aware of the study conditions. Stanley Schachter and Jerome Singer used a double-blind design in an early study of physiological determinants of emotional states (Schachter and Singer 1962). Schachter and Singer told subjects they were studying how vitamin injections affected vision. Subjects were asked if they would agree to be injected with "Suproxin" and 95 percent of them agreed. After injection, they were put in a waiting room with another subject and told to wait for 20 minutes until the vitamin took effect.

Schachter and Singer were actually studying how social situations affect subjects' emotional states after arousal. Subjects were actually injected with either epinephrine (adrenaline) or a placebo. Adrenaline produces a state of physiological arousal and subjects were given (1) no information about how the injection would affect them, (2) given accurate information about how the injection would affect them, or (3) given false information about how the injection would affect them. The second "subject" in the room was a confederate who had been told to behave (1) playfully (the euphoria treatment) or (2) angrily. The subjects were unaware that researchers were studying the emotions they would describe after interacting with the confederate. The confederate did not know if the subject had received epinephrine or a placebo or what experimenters had said about the aftereffects of injection. Double-blind designs reduce the possibility of experimenter (or subject) bias substantially.

An alternate approach to experimenter bias is to study the effects of biases deliberately introduced. Willer deliberately introduced experimenter biases into exchange network studies and measured their effects on the end conditions, exchange ratios. The structure studied was like the strong structure studied by Brennan. Through exclusion, exchange ratios should move rapidly toward the extreme favoring the central position. The assistant conducting the experiment was given a series of negative cues to be employed contingent on the exchange ratios agreed on by subjects. As those ratios moved *toward* predicted values, the assistant hung his head and began mumbling, "Now I am in trouble, big trouble." If the process continued he said, "This is a disaster." And later, "This is it. I am going to lose my job." Corresponding positive cues were developed to be voiced if data moved *away* from predicted values.

It is useful to know whether an experiment is sensitive to experimenter bias and, if so, its degree of sensitivity. Deliberately introducing experimenter bias has the advantage of giving a quantitative value to the vulnerability of the experiment to biases. As reported in Willer (1987:198ff.), the design just discussed was almost completely immune to experimenter bias. The cues slightly slowed the power process, which nevertheless eventually approached the predicted extreme. Only when subjects were no longer paid by points earned in exchange, but, instead, were paid to "please the experimenter" was the power process halted and to some degree reversed.

Detecting and Reducing Artifacts

Researchers can reduce demand characteristics and experimenter bias by carefully designing experiments with an eye toward potential difficulties. They can create procedures for detecting potential sources of artifacts and build them into their designs; and they can rigorously pretest experimental designs. Experimental researchers usually conduct pretests, systematic trial runs, of their settings before they begin the study. It is common practice to draw persons from the subject pool and run them through what are, in fact, full dress rehearsals. Pretests can be stopped at any point and the "subjects" can be questioned and debriefed. We use as an example, the earliest designs for a series of studies one of us conducted.

Walker and Zelditch (1993) have conducted many experiments that study the responses of individuals to structurally based economic inequality. In one series of studies, subjects work at a puzzle-solving task. A group of five subjects is given 10 puzzles, one on each of 10 trials. Each subject is given some pieces of information necessary to complete the puzzle. However, no subject can complete the puzzle without getting the information held by every other group member. The group of subjects is paid for every correct puzzle solution (i.e., a maximum of five correct solutions for each puzzle), and the earnings are equally divided among the five.

Some groups begin work in the "wheel structure" shown in Figure 5.1, but group members are told that a majority (3) can change the shape of the structure, and are also told how to change the structure. In the earliest studies, subjects believe they are sending and receiving puzzle information by passing written messages among themselves but the messages are actually handled by the research staff. After working several practice puzzles to become familiar with the system, the experimenter created an incentive for group members to

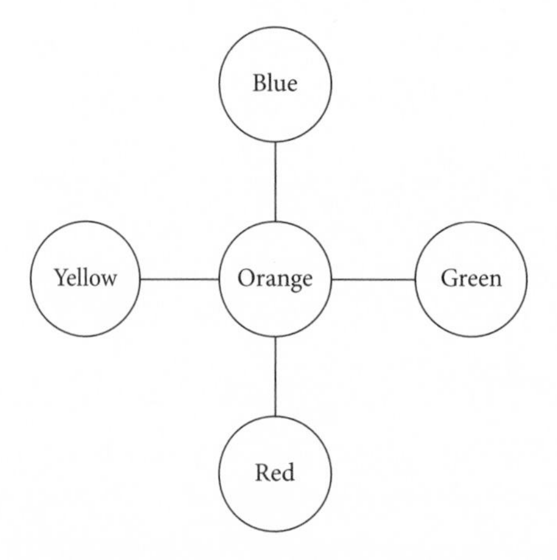

FIGURE 5.1. A Centralized "Wheel" Structure

work faster. He gave a bonus payment to the member who turned in the first correct solution for each of the 10 puzzles. The bonus created competition and the initially cooperative task became a mixed-motive, cooperative-competitive task. Although the centralized wheel structure is the most efficient structure for this kind of communication, that structure ensured that the central actor (Orange in Figure 5.1) would win *all* bonuses. The gross inequality created pressure for peripheral actors to try to change the structure. One set of studies explored the central actor's reactions to attempts to change the structure.

Walker and Zelditch, using theories from political sociology and political science, formed the hypothesis that the central actors would resist changes to the structure because change would dilute the *power* they had to control interaction *and* it worked against their interests as rational actors. Change would reduce their earnings substantially. Pretests of the setting with subjects in the central role showed that almost all did resist attempts to change the structure. The data also showed that they often failed to turn in their answers first and, hence, "lost" the bonus on many trials.

Post-session interviews revealed that the subjects' motives were very different than the investigators' hypothesized. Central subjects resisted change because (1) they recognized the structurally based inequality and thought it unfair, but (2) they feared that change would permit some other member to gain control and "hog" the bonuses. Their solution was to resist change and retain power so that they could ensure that everyone got a fair share of bonuses. As central in the information network, they could decide *which* other member first received all of the puzzle information, could wait for that member to turn in his information, and then give the information to the remaining members so that all could get the correct solution. Their solution of "rotating" bonuses ensured that they and every other member got nearly equal shares of the payoffs. Without pretesting and post-session interviewing, Walker and Zelditch would not have learned the subjects' motivations nor could they have redesigned the experiments to remove equity and justice considerations.

Lab Settings and Computer-Mediated Experiments

This section will introduce you to two very different computer-mediated settings for experimental research in sociology. Both are "experimental instruments" used in sociology much like a linear accelerator and a mass spectrometer are used in physics. That is, they generate experimental conditions and measure results. Neither will be entirely new to you because they are instruments for experiments you have seen in the previous chapter, experiments that test Status Characteristics Theory and Elementary Theory. For both theories, the goal of the instrumentation is to increase experimenter control over the setting encountered by the subject beyond that possible with earlier setups. To the extent that the goal is attained, (1) theoretic models and experimental replicas become increasingly similar, (2) both experimenter and subject biases are reduced, and (3) replications can be identical to initial runs. We explain how you can access these instruments for your research.

The advent of computer-mediated experimentation has fundamentally changed the appearance of the laboratories in which sociological research is carried out. For example, the lab at the University of Kansas, that one of us used some years ago, was an office temporarily set aside for conducting experiments. It contained chairs, piles of poker chips to serve as counters for negotiations, and an egg timer used by the experimenter to time negotiation rounds. Network shapes were built by placing plywood partitions to block some interactions: those interactions not blocked formed the experimental network. In fact, the setup of that lab was described as part of the discussion of the Brennan experiments in Chapter 4.

The Laboratory for Sociological Research at the University of South Carolina, where we carry out our collaborative research, exemplifies well the layout of a laboratory setup for computer-mediated experiments. It is both larger and fundamentally different from the Kansas lab. As seen in Figure 5.2, there are twelve small rooms, including two control stations, around the periphery of the larger central room. Each small room contains a PC while the large room contains eight more PCs on each side. Experiments place subjects in the small rooms where no inter-subject contact beyond that mediated by the experimental instrument is possible. For larger networks, subjects may also be placed in the two small rooms labeled in the figure "Control Stations" with control of the experiment moved to one of the PCs in the large room.

We now turn to the two experimental instruments and, in doing so, we ask you to keep foremost in mind that experiments that investigate theory should be designed by theory. In fact, each of the experimental instruments now to be described was designed by the theory it tests. You will notice similarities between the setup of the Moore experiment and the instrument built to test Status Characteristics Theory. You will also find similarities between Brennan's setup and the instrumentation built to test Elementary Theory. Why these similarities? The answer is not because the electronic instrumentation

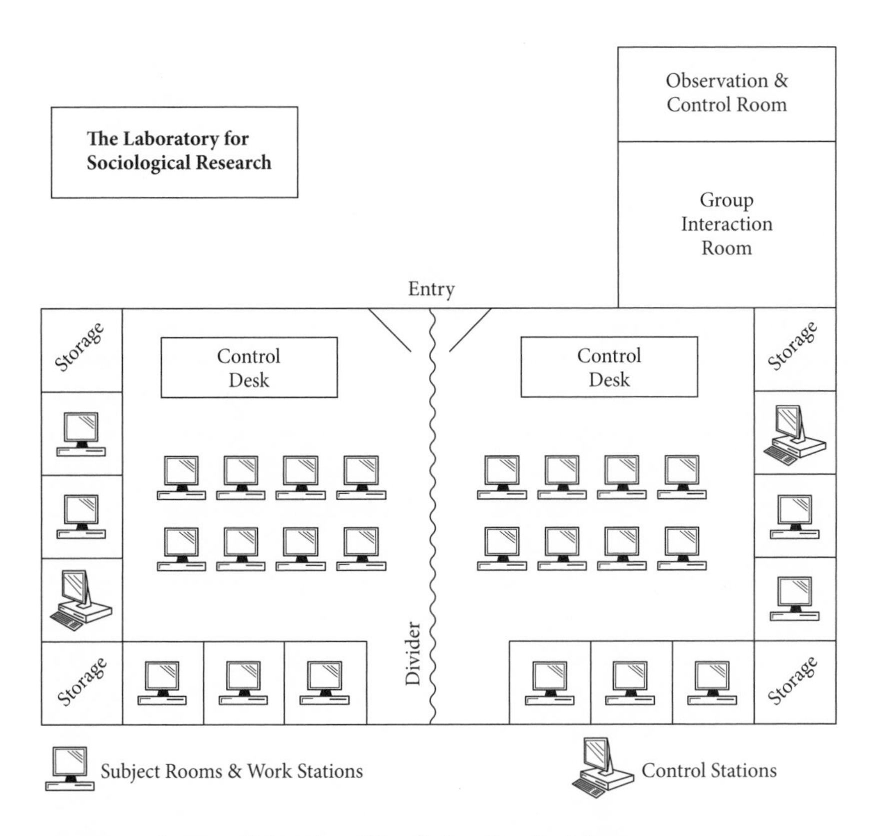

FIGURE 5.2. Layout, University of South Carolina Sociology Research Laboratory

was designed to copy the setup used in the earlier experiments. Instead, theory produces the similarity. The earlier setup and the electronic instrumentation for each theory are similar because both were built in light of its theory.

An Electronic Instrument for Status Characteristics Experiments

Lisa Troyer of the University of Iowa developed the instrumentation we now describe. She makes it available to you and other interested scholars without cost. Nevertheless, she asks those with funded projects to make a donation to the Center for Group Process at the University of Iowa. Donations are used to support student research projects.

With this software, it is possible to carry out new Status Characteristics Theory (SCT) experiments or to replicate Moore's (1968, see Chapter 4) and many other SCT experiments. Initial conditions for the electronic experiment are set in much the same way as did Moore. It will be remembered that tests of SCT require situations in which actors differing on one or more status characteristics use status information to organize their interaction. The subject is seated at a PC that provides a complete set of task instructions. To produce the collective task orientation, the instrument informs the subject (1) that she will be working with a partner, (2) that the objective is to give the best responses to the questions that will be posed, and (3) to that end she will be given information on the responses of her partner.

The subject also receives information that permits comparison of self and other on some status dimension. For example, if a 21-year-old undergraduate with a C+ average is to be low status, she could be told that her partner is a 30-year-old graduate student with an A average. For added experimental realism, the subject could be given a snapshot of her partner chosen to represent a high or low status individual. Instructions assure subjects that they will not meet their partner after the study.

Like Moore's study, each experiment contains one real subject and one fictive subject, but here the role of fictive subject is played, not by the experimenter but by the computer. The instrument presents the Contrast Sensitivity Task to the subject. The subject views a series of slides on her PC like that shown in Figure 5.3. As can be seen, the two rectangles, one above the other, have nearly equal irregular patterns of black and white. Below the slides are three rectangular panels each with a "light" and a button on each side. The first panel has the label "Your Initial Choice" in the center. The leftmost button is labeled "top" and the right button is labeled "bottom." The second panel is labeled "Your Partner's Initial Choice" and the third panel is labeled "Your Final Choice." The left and right buttons on the last two panels are also labeled "top" and "bottom."

The slide appears for approximately 5 seconds and then, using mouse control, the subject makes an initial judgment as to whether there is more white (or black) on the top or bottom rectangle by clicking one of the two buttons on the initial choice panel. The appropriate light illuminates and there is a delay after which the PC illuminates a light representing the choice made by the subject's fictive partner. The delay after the subject's initial choice is computer-mediated to vary somewhat reflecting varying times needed for decision by the fictive other. Then the slide reappears and the screen instructs the subject to record a final choice on the third or bottommost panel.

The PC's program ensures that the subject and her ostensible partner disagree on the number of trials set by the experimenter. As in Moore's experiment, the difference between subject's initial and final choice when the

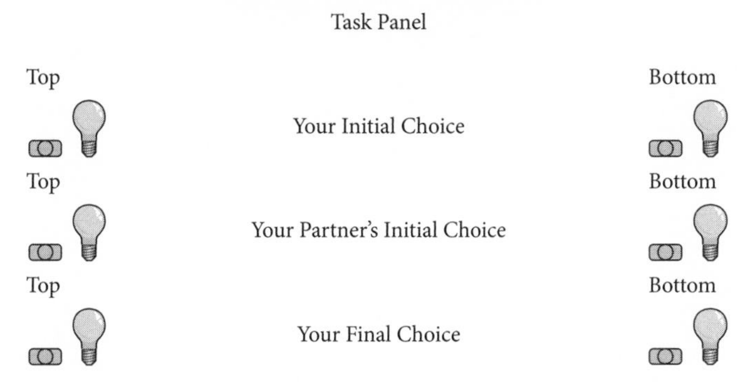

FIGURE 5.3. Two-Slide Contrast Sensitivity Task for SCT Experiments

latter is consistent with the choice of the fictive other is the measure of influence. The frequency that the subject stays with her first choice on criterion trials is p(S), the probability of a stay or self response. As before, high p(S) values indicate greater resistance to influence and the converse for low values of p(S). SCT implies that high status actors will have higher p(S) values when interacting with low status partners who are expected to have lower values of p(S). The setting permits a researcher to investigate the effects of a wide range of diffuse (e.g., gender, sexual orientation, educational status) and specific (e.g., musical ability, athletic ability, or completely fictional abilities) characteristics.

What kind of physical layout is needed for this SCT instrument? Remembering that each experiment has only one real subject, very little is needed. Still the physical arrangement should be such that the subject will find it plausible that there is also another subject participating in the experiment. One way to generate that plausibility is to have a layout where two subjects actually can be run simultaneously. The USC Laboratory for Sociological Research, the layout of which was shown above, could be used for these experiments.

But such a large lab is perhaps overkill because, using all the small rooms around the periphery, as many as 12 studies could be run simultaneously.

An Electronic Instrument for Elementary Theory Experiments

Unlike SCT experiments, all sessions of Elementary Theory (ET) experiments have multiple subjects ranging from as few as two to 10 or more. The experimental instrumentation, ExNet allows subjects to interact with each other using the PC as an interface. Located at http://weblab.ship.edu, the ExNet experimental system is used online and links subject computers through the Internet. Each study is set up by the experimenter who accesses and activates the software through the Web browsers on her PC and those the subjects use.

ExNet is designed to test ET: the replicas it builds are intended to correspond to models built by ET. (It can also be used to test other theories in which social relations, such as exchange, occur in structures.) ET models actors as rational decision makers. If subjects are to correspond to theoretic actors, each must have effectively full information about the structure in which they are interacting. To that end, each subject's screen will display the network, indicate where that subject is located, and show all activity including offers, counteroffers, and exchanges as completed by all subjects in the structure. Alternatively, information displayed to each subject can be restricted in a number of ways.

Shown in Figure 5.4 are three such structures. Each subject can easily locate her own position because its letter is larger than the others. All three structures of Figure 5.4 are displayed from the point of view of a subject located in B. The screen of the A subject will, of course, show her letter largest and similarly for C's screen. If subjects are to be rational, they must master the subject-PC interface. Thus, each session is preceded by a tutorial explaining how to make offers, read the offers from others, and complete exchanges. To assure expertise on the part of subjects, just prior to the experiment, they practice on a structure, distinct from the one under investigation.

To test the full range of structures modeled by ET, ExNet must generate an array of structural conditions. In fact, ExNet can investigate all structural

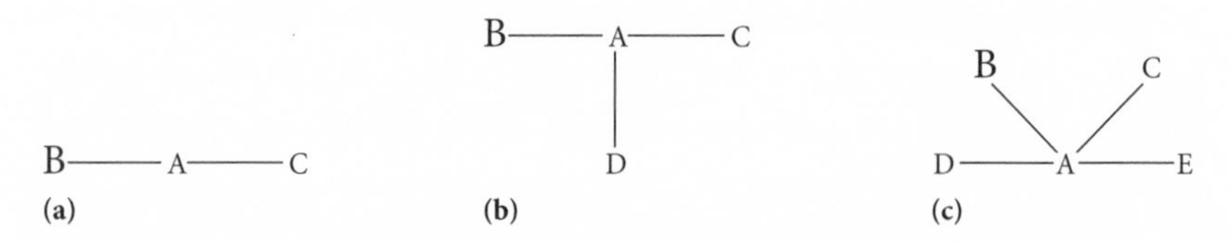

FIGURE 5.4. Subject View of Three Network Exchange Structures

properties thus far discovered by ET in any structural configuration for up to 25 subject positions.

Each ExNet session will take the following form. As subjects arrive, they are shown to the small rooms at the periphery of the lab (see Figure 5.2), each of which contains a PC that is already online and connected to ExNet. The tutorial is started, followed by a practice structure, and then the experimental structure. In the experiment, subjects negotiate though a series of rounds, each of which begins with the resources renewed. Each round ends when all exchanges that can be completed are completed or, failing that, when a given time limit is reached. The experimenter can group rounds into periods and electronically rotate subjects across positions at the beginning of each period. For example, each subject can be placed in each structural position for one period.

Providing each subject with full information produces very high experimental realism. High subject involvement is frequently observed as subjects, though alone, vocalize: "Why wasn't my offer taken!" and "No, that offer is too small!" On rare occasions, very excitable subjects have been known to bang on walls.

A number of design qualities are purposefully built into ExNet. First and foremost, ExNet's replicas are typically an excellent fit to the theoretic model being investigated. For example, the structures of Figure 5.4 are network configurations for models drawn from the theory and for replica structures as displayed on the subjects' screens. No communications can occur in the experimental replica other than those modeled in the theory and realized in the replica. In the figure, B cannot communicate with C, nor can C communicate with D, and similarly for all positions not immediately connected to each other. Furthermore, the content of communications is controlled. Each subject is free to offer any resource division to anyone to whom she is connected, but no threats or side payments can be sent. These high levels of control stand in considerable contrast to the array of communications beyond those modeled that could and did occur in Brennan's face-to-face experiments.

ExNet allows a number of other important theoretical properties to be studied in quite pure form. Subjects' decision making may or may not actually correspond to the rational decision making of ET's models. Nevertheless, conditions necessary for rationality are present including full information, well-trained and thus competent subjects, and the high levels of involvement needed to maintain attention. Qualities idiosyncratic to one or another subject such as better negotiation skills can be canceled out by rotating subjects across structural positions. Because subjects occupy both advantaged and disadvantaged positions, rotation has the added benefit of producing greater equality of subjects' earnings thus dampening equity concerns.

Not all universities have dedicated labs and, when they do not, ExNet experiments can be run in a computer classroom. A computer classroom can be used if the computers are connected to the Internet and have Internet Explorer. Of course, a computer classroom will not provide the levels of subject-to-subject isolation possible in a dedicated lab, but in some cases that will matter little. Subjects interacting through ExNet are very busy reading their screens and clicking icons to enter their own offers, counteroffers, and agreements. They have little or no time to attend to others in the room. Furthermore, that subjects can see each other may be less important than it first seems. With the dyad as an exception, it is impossible for any subject to know which of the others is in what network position.

Summing Up

Across the sciences, it has been shown repeatedly that linking experiment and theory is the best way for precise and general knowledge to grow. The recent explosive growth of knowledge in experimental economics shows what can be accomplished. Closer to home, over the last two decades in sociology, experimentally grounded theory has grown and, beyond growth, accumulated. If experimental research now grows in sociology as it has in economics, in a decade theoretically driven experimentation would be the leading method of our field.

We now turn to the concluding chapter of this book and take up three significant issues. First, we examine the uses of the experimental method in sociology to show that those uses are much broader than previously thought. Experimental research is not confined to the study of small groups. We are lifelong experimentalists and almost all of our work has aimed at the explanation of large social structures. Second, we ask what kind of research methods should be employed when theory cannot be tested experimentally. We explain the idea of scope sampling and show how theoretically driven historical-comparative investigations can take a form logically like the theoretically driven experiment. Third, we ask whether that kind of investigation and experiments can be linked.

6 THE FORMS OF CONTROLLED INVESTIGATION

THIS CHAPTER EXAMINES MODES OF INVESTIGATION THAT can be employed in sociology to test and apply theory. We call those modes of investigation "controlled investigations"—a term we borrow from the philosopher of science, Ernest Nagel (1961). The term "controlled" in controlled investigation refers not to experimental control but to the control of investigation by theory. Thus theory-driven experiments are controlled investigations, but so are certain other kinds of research that have little or no experimental control. In fact, all controlled investigations are logically like theory-driven experiments, and, because they are, their results carry the same weight even when the investigator controls neither initial conditions nor the conditions under which results are observed.

Before we take up other forms of controlled investigation, we respond to claims that the uses of and implications of social science experiments are very limited. The belief that social science experiments have limited utility is not new. Max Weber, in a rare reference to the experiment, asserted that it could be employed only "in the few very special cases" ([1918] 1968:10). Later Burgess (1929) asserted that, unlike physicists, sociologists cannot bring their phenomena into the lab. "The objects of social science research, as persons, groups, and institutions, must be studied if at all in the laboratory of community life."(1929: 47). More recently, Lieberson (1985; Lieberson and Lynn 2002) has argued that the experiment is of little or no use in sociology and, more perniciously, that a methodology modeled on the experiment as practiced in physics is inappropriate to sociology (1985 passim).

These assertions grow out of fundamental misunderstandings of science and the place of theory in experimental and other research. For example, Burgess is wrong in thinking that the key to the success of experimentation in other fields is the ability to bring objects from outside the lab into it. To the contrary, the key to success is theory, the application of which bridges from the laboratory to the field. To Lieberson, "A key feature of the experimental approach is that subjects are randomly assigned to the conditions under study" (1985:14) and that this method is "derived from classical physics, a model that is totally inappropriate for sociology" (Lieberson and Lynn 2002). Readers of this volume know that Lieberson is misinformed. It is difficult to imagine what things classical physics would randomly assign or what would be gained by that assignment. Whereas Fisher introduced random assignment in 1935, three hundred years earlier in physics, the theory-driven experiment was already well developed by Galileo. Below we will show that Weber was wrong in thinking that the uses of the experiment are highly limited. In fact, some experiments help explain macro-structures central to his theoretic. Yet Weber was not wrongheaded. Believing the experiment was limited, he proposed the "mental experiment," a kind of simulation built on the logical structure of the experiment ([1918] 1968:10).

This chapter has three sections. First we explore the utility of the experiment and find that it is far greater than its critics believe. Next we look to nonexperimental forms of controlled investigation and explain their design. Finally, we conclude the chapter and the book by asking whether experimental and nonexperimental modes of controlled investigation can work together and whether they should now be working together to advance the field of sociology.

The Utility of Theory-Driven Experiments

Sociology experiments are presumed to have limited utility for three reasons:

1. Experiments study only small structures and their time span is brief. How can their results be brought to bear on large, long-standing structures?
2. Experiments are simple. How can they be brought to bear on complex phenomena?
3. Experiments lack external validity. How can they produce general knowledge?

As we now show, none of these limit the utility of the experimental method.

Is Scale a Problem?

Issues of scale are frequently cited as reasons why the experiment is of limited utility to sociology. For example, Nagel proposed the idea of controlled investigations because social scientists cannot experiment on large-scale and long-term processes, such as the rise of industrial capitalism (1961:451–2). But is it true that experiments can say nothing about the rise of industrial capitalism? Here we will not try to encompass the whole of that process, but point to a social structure that many, including Weber, hold was a necessary part of it because it impacted legal systems. According to Weber, industry cannot develop if it is subject to an arbitrary, confiscatory legal system.

A legal system compatible with capitalism could be established if the burghers of late feudal era cities were empowered to do so. The key was to gain autonomy for their cities. In Europe during that era, cities gained autonomy by playing off king against local lord, and both against church powers. Autonomy for cities included reduced taxes and, for the burghers, the right to write and enforce their own laws. Said differently, cities auctioned the right to coerce them to the powerful bidder making the best offer. King, local lord, and church were competitors because, due to the decentralization of feudalism, their sovereignties overlapped.

Later, after the fall of feudalism, power centralized into state structures, but the competition among coercive powers did not end. By moving or threatening to move investments across state boundaries, firms with mobile capital played states against each other, a game that continues today. Today the game is played, not just among national states, but among states within the United States. For example, the state of South Carolina has a well-funded office, the purpose of which is to develop and make offers to corporations to induce them to move there. Frequently that office is outbid by corresponding offices of other states or countries.[1]

Figure 6.1 shows a coercee-central structure. In the structure, D must select one of three Cs to be its coercer, but may negotiate with all three before that choice. No C can transmit its negative sanction before reaching an agreement with D. No further parameters need be specified to make three important predictions. First, the Cs will compete to be D's sole coercer. Second, no C will ever transmit its negative sanction. Third, because of competition, the rate of coercive exploitation will approach the minimum possible for the system. In the field, low rates of coercive exploitation mean low taxes and opportunities for autonomy. In fact, two of three coercers will be excluded, and experience with exclusively connected exchange structures like those studied

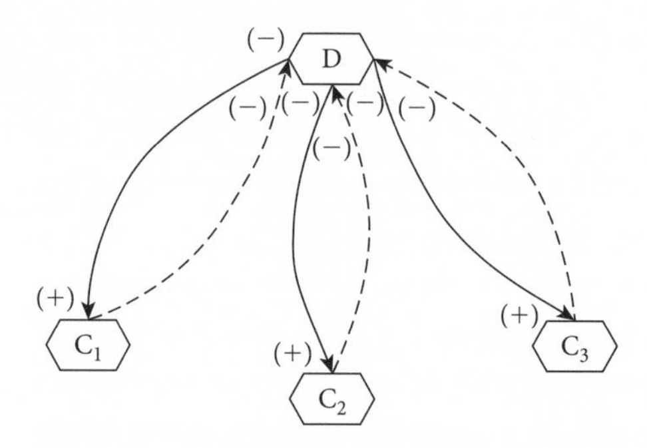

FIGURE 6.1. A Coercee-Centered Exchange Structure

by Brennan leaves little doubt that the three predictions will be supported. Cities playing off feudal powers and mobile capitalist firms playing off states are two examples of coercee-central systems now theoretically modeled. All that remains to link experiments to these historic structures is to carry out the five steps outlined in Chapter 4, building the replica for the coercee-central model and conducting the study.

What evidence is there that competition among coercive powers is a necessary condition for the rise of capitalism? According to Weber, capitalism did not develop in classical China, despite (1) its technological superiority to Europe and (2) the healthy acquisitive motives of the Chinese. In classical China, commercial law remained undeveloped and state confiscation of the assets of developing capitalist firms was a widespread practice. Why was China's commercial law not compatible with the rise of capitalism? Because China was a single unitary empire. Therefore, unlike Europe, there were no competing powers to play against each other. Thus there was no urban autonomy and no compatible legal system (Weber [1911] 1951). Here again there is a relevant model, coercion without exclusion, the experimental results of which could be compared to the coercee-central structure described above.

Perhaps a second example is needed to drive home the point that scale need not be an impediment. If Nagel is right about capitalism, it also follows that experiments can say nothing about the rise and fall of the great slave structures of antiquity and the early modern Americas. Contrary to Nagel, experiments on coercive structures have been run and they support Marx and Weber's insights on slave structures. Marx ([1876] 1976:377) and Weber ([1896] 1976:398) linked rates of coercive exploitation—that is to say profit—to the supply of cheap slaves. In the nineteenth century, slavery in Jamaica and Brazil, where slave importation was unrestricted, was much more profitable than in the United States where importation was contrary to law. Their insights were historically sound, but neither offered any reasons why profit and slave importation were linked.

Elementary Theory predicts that rates of coercive exploitation should vary with the slave owner's cost of sanctioning slaves. In turn, sanctioning costs are associated with the supply of cheap slaves. As sanctioning costs decline, there is a point at which slaves compete to avoid sanctioning. Competition benefiting the coercer is like the competition benefiting central positions in the strong exchange structures Brennan studied (1981). That competition results in very high if not maximum rates of coercive exploitation. Experiments testing that derivation offered strong support for it (Willer 1987).

It is difficult to imagine structures more different in scale than the gigantic slave estates of antiquity or the large slave-based latifundia of the eighteenth- and nineteenth-century Americas and the small experimental coercive structures that test derivations from Elementary Theory. Such differences are problematic to those who fail to understand the logic of theory-driven research. Experimental tests of theory do not require scientists—whether sociologists or astrophysicists—to reproduce scale models in their labs. What is needed is something very different—a structural similarity between lab and field. Coercee-centered structures like those we have just described look very similar to slave structures when they are *seen from the point of view of theory.*

Beyond scale there is a second important difference between the historical cases and the experimental replicas just discussed. The historical cases are much, much more complex. What is the justification for connecting the results of simple experiments to the complexities of history?

Theoretic Simplicity and Complex Structures

Theory can link simple experimental replicas to complex historical and contemporary social structures because it has an important capability not yet discussed in this book. *Theory can compose complex models.* The simple models this book has used in building experiments can be combined and recombined to compose increasingly complex theoretic models. The composition of more complex formulations out of simpler ones was already seen on a smaller scale when social relations were composed into structures. By extension, multiple structural models can be built into a single complex model. That complex model can be used for explanations in the field. Importantly, it can also be used to build large and complex experimental replicas or simulations. Regrettably, the method of composition, though receiving passing mention (Willer 1984), is almost unknown today in any social science. Its absence is odd because it is well known in other sciences (Einstein [1934] 1954:270ff). Furthermore, in the social sciences, the compositional method was proposed as early as Marx ([1857] 1973:100ff).

Is the compositional method feasible? Is social theory advanced enough to link simpler models and draw new conclusions? Here is a hypothetical example. Shown in Figure 6.2 are two linked exchange structures α and β with C, the "capitalist," occupying the central position in both. C is like a modern capitalist with mobile capital. That is, C owns all the exchange relations and all the positions in both networks and can freely move relations across the two structures from α to β or the reverse. The α structure is null connected such that, if β did not exist, exchange would occur at equipower. In the absence of β, wages would be high and profits low. Being null connected, α is not itself a power structure favoring the capitalist.

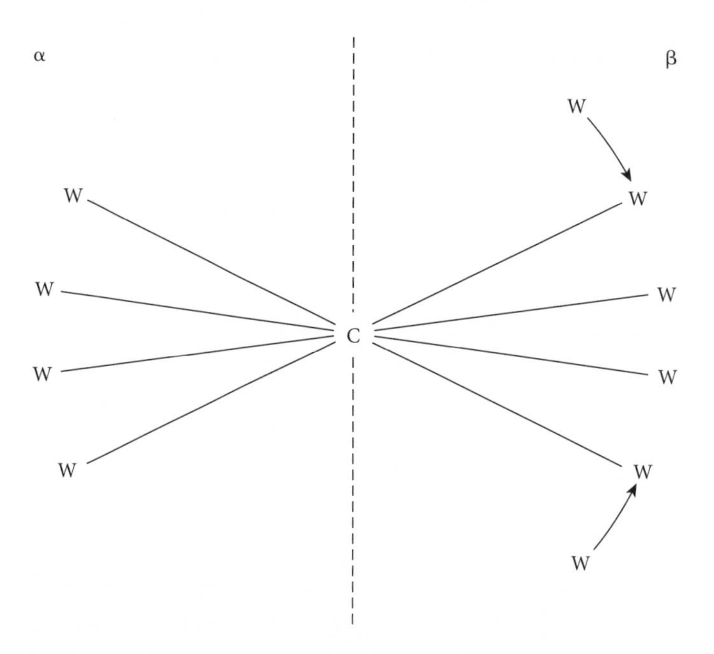

FIGURE 6.2. Model of Two Linked Exchange Structures

Now consider the β structure. It is strong power with a surplus of Ws competing for C's positions. Let there be a surplus of Ws whatever the number of relations moved from α. Because β is a strong-power exchange structure, its exchange ratio will maximally favor C. That is, profits will be maximal and wages minimal. A new inference immediately follows. C has a credible threat to move relations from α to β and a rational self-interested C will do so unless the exchange ratio in α becomes as favorable to C as the exchange ratio in β. Because only α and β structures exist in the model, Ws in α have no alternative jobs and will accept the minimal wages.

Thus, in complex networks where null and exclusively connected networks are linked, the model sketched above asserts that null-connected networks become power structures. Once formally stated, an experimental replica could be built to test the model. Furthermore, there are links outside the lab. In the world economic system today, capital is highly mobile. Job transfers from the United States to Third World countries do threaten wages in the United States.

Is research that builds and applies increasingly complex models feasible today? We suggest that it is because theoretically more difficult work is already ongoing. The complex model just discussed is a straightforward extension of a single theory, Elementary Theory. More difficult is research that bridges across two theories. At least two bridging research projects are now under way. Status Characteristics Theory and Elementary Theory are linked in two ways by Thye, Willer, and Markovsky to relate influence processes and the spread of status value to exchange ratios (2006). Experimental research we now have under way links Legitimacy Theory and Elementary Theory showing that levels of legitimacy affect whether power can be countervailed.

Theory, Experiment, and External Validity

Some claim that the experiment is of little utility because it does not have external validity. What is "external validity" and why do some believe it to be important, important enough to reject the experiment?

By "external validity" critics mean the ability to infer a population parameter from a sample randomly drawn. (See Lucas 2003 for an extended discussion.) It is true that experiments are of little or no use in inferring population parameters. But such inferences are not wanted or needed. The objective of theory-driven experiments is to test theory that is general and universal. By "universal" we mean that theory applies without regard to special conditions of time and place. Theories must claim universality if they are to have explanatory and predictive power. By contrast, any sample with external validity allows inference only to a population at a specific time and specific place. Because inferences are to specifics and not to general phenomena, neither explanation nor prediction follows from them.

The laws of theory are not population estimates. The scope of sociological theory, like all theory, cannot be defined by an enumerable collectivity, such as a population of people. Unlike the collectivities sampled in surveys, the numbers included in any theory's scope cannot be counted, cannot even be listed. Instead, the scope of theory refers to a universe of phenomena. Furthermore, that scope is not fixed. As the theory is tested, evidence may find that its scope is narrower than previously thought. Alternatively, as the theory grows and new formulations are introduced, the scope of the theory expands.

It immediately follows that Popper is right (1994). Unlike inference from sample to population, theory testing is not inductive nor can it be. For induction, surveys rely on a list of instances, usually people, from which a random sample is drawn. By contrast, the universe of applications of theory cannot be counted and cannot be listed. Thus random sampling is impossible and, because it is, so is induction. Instead of induction, in experiments the direction of inference is reversed. In theory testing we do not induce from experiments to theory, but deduce from theory to experiments.

Nonexperimental Forms of Controlled Investigation

Two questions remain. The first question concerns the empirical cases outside the lab used to test theory. In experiments, the empirical cases that test theory are replicas created in the lab. In nonexperimental controlled investigation, the cases that test theory must be found. More specifically they must be selected. What should govern that selection? We propose "scope sampling" as the answer. Second, how should each investigation be designed and carried out? We propose the same answer as given to the question, "How should experiments be designed?" Theory should also design and guide nonexperimental research.

Systematic selection of empirical cases for theory testing is called "scope sampling." This procedure can most easily be understood by analogy from the theory-driven experiment. In theory-driven experiments, we construct empirical cases that are replicas. The replicas are arrayed along variations of the theory's initial conditions. For example, experiments in Status Characteristics Theory (SCT) routinely investigate high status people who believe they are interacting with low status people and low status people who believe they are interacting with high status people. The initial conditions of these experiments are arrayed along a *scope* of varying statuses from high to low. Similarly, initial conditions of Elementary Theory (ET) experiments will array replicas along a *scope* of varying structural conditions, a *scope* of varying structural configurations, or both.[2] The Brennan experiments varied structural conditions from null to exclusion. Exchange experiments in Chapter 1 varied configuration from triangle to centralized network.

There are two equally important reasons for arraying replicas along the scope of initial conditions of the theory. First, that array offers an excellent test of the theory's ability to link initial and end conditions. For SCT we want to assert that status differences produce influence: arraying replicas by status can support that statement. For ET we want to say that power varies with structural conditions and configurations: arraying replicas by those two conditions can support that statement. Second, arraying initial conditions by scope is the best test for the theory because it maps out the confirmed scope of the theory, the range in which it is known to predict effectively and explain. Furthermore, it is the most rigorous test, because it maximizes the opportunity for error and thus falsification.

Scope *sampling* in nonexperimental research follows exactly the same logic as varying scope in the experiment. When a scope is sampled, instances are selected following variations in initial conditions of the theory's models. Here is an example. Drawing on the previous section, models for coercee-central structures could be tested using historical cases. In China's unitary empire, coercive relations were everywhere, but there were no coercee-central structures. Sovereignties did not overlap and there were no small states between which capital could move. In China high rates of coercive exploitation should be found. By contrast, Europe had coercee-central structures, first due to overlapping sovereignties and later due to the ease of moving capital between Europe's small states. Low rates of coercive exploitation should be found in Europe. Two measures of coercive exploitation levels are tax rates and the fit of the legal systems to the needs of business.

Consider a second example that applies SCT and ET together. Again we array cases by the theoretic scope—in this case by the status of buyers in a market. Is status inversely related to price paid for commodities as SCT and

ET together predict? For example, do black women pay more for the same car than white men? Recent evidence suggests that black women do pay more (Ayres 2006). Thus the SCT and ET linkage is supported.

The aim of traditional contemporary-comparative and historical-comparative studies has been to flush out empirical regularities using Mill's methods such as difference, agreement, and concomitant variation. Because Mill's methods provide no criteria for distinguishing things that should be observed from things that do not need observation, traditional comparative research is open-ended. There is no point at which it can be rationally concluded. For example, study cannot be concluded by successful discovery of regularities, for experience has shown that the methods do not uncover the absolute regularities Mill expected. The best comparative studies in sociology, like the work of Prechel (2000) and Mann (1986), do not limit themselves to Mill's methods. They overcome Mill's shortcomings precisely because they go well beyond Mill and are guided by formulations from theorists like Marx and Weber. Those formulations guide researchers to what should and should not be observed and to what relations might and might not be found. In so doing, they allow comparative research to be rationally carried out and rationally concluded.

Nevertheless, nonexperimental applications of the more formal theory of the kind used in theory-driven experimentation improves on research guided by classical theory and gives a new kind of contemporary-comparative and historical-comparative study. For contemporary research, formal theory directs the researcher to what should and what should not be observed. For historical research, formal theory directs the researcher to the kinds of data that need to be unearthed to apply the theory. Working back and forth between empirical case and theory, first simple models are constructed that are then composed into the increasingly complex formulations needed to cover the case. In fact, examples of models that apply to many historical cases were given in outline form in the previous section including one model composed of two linked structures.

Just as experimental tests seek to cover substantial scope, scope sampling is a rigorous test for the theory because it maximizes the opportunity for error and thus falsification. And, like experiments, scope sampling broadly maps out the confirmed scope of the theory, the range in which it is known to predict effectively and explain. Importantly, the logic of theory-driven experimentation and scope sampling is the same, and, as a result, their tests of theory carry the same weight.

Connecting Experimental and Nonexperimental Investigation

Can experimental and nonexperimental controlled investigations work together to offer predictions and explanations of greater precision and scope than previously possible? This chapter has already shown that both kinds of investigations are driven by the same logic, can even be driven by the same theory. Not only is the goal of combining the two kinds of study feasible, but the first steps toward that goal have already been taken—but examples are few. Beyond Cohen discussed in a Chapter 4, Willer and associates (1997) applied experimentally tested models to Weber's interpretation of the fall of ancient civilization in the West. Models central to that application included the two coercive structures applied to slavery used earlier in this chapter. Bell, Walker, and Willer (2000) applied models from Elementary Theory, Status Characteristics Theory, and Legitimacy Theory to formal organizational structures to explain the very high levels of obedience found there.

Understanding of the physical universe has exploded over the last hundred years because physics and astronomy have combined experimental and nonexperimental research. Understanding of the animate universe has exploded over the last 50 years because genetics and evolutionary biology have combined experimental and nonexperimental research. Over the next 50 years understanding of the social universe will explode. That explosion will begin when social theory routinely combines experimental and nonexperimental research.

Endnotes

Notes to Chapter 1

1. For similar definitions see Campbell and Stanley (1966:1) or the longer but consistent definition in Selltiz et al. (1959:94).

2. Bales' contributions to sociological theory are discussed in his many books and articles including *Interaction Process Analysis* (Bales 1950) and *Social Interaction Systems* (Bales 1999). According to Bales (1999), his early interests in and insights about elementary group processes grew out of his observations of naturally occurring groups like faculty meetings.

3. A close parallel to the study discussed here is Skvoretz and Willer (1991).

4. Stipends were pegged to unemployment benefits in the respective states. Inmates released from Georgia prisons received $70. Those released in Georgia could earn an additional $13.75 from wages before their stipends were reduced. Inmates released from Texas prisons got $63 per week and could earn an additional $8 before their benefits were reduced.

5. Just as theory-driven experiments were done before Galileo, empiricist experiments predate Mill. Bacon ([1620] 2000)—Galileo's contemporary—expressed ideas that presage Mill. Yet, Mill's systematic statement on the issue is more complete and has had greater influence on current scientific practice.

6. Consider the population of all students at X University that are enrolled in introductory psychology courses that require students to participate in any three of five possible experiments. Our experience suggests that most will be freshmen or sophomores and that disproportionately few of them will be physics or engineering majors.

Notes to Chapter 2

1. We use lowercase symbols like *x* or *y* to represent concrete events or phenomena. Susie's marriage to John is a specific event. We use uppercase symbols (e.g., *X* and *Y*) to stand for event-classes. Event-classes are aggregations of events of the same type such as, for example, all marriages sanctioned by the states of South Carolina and Arizona in 2006. Following Willer and Webster (1970), we use these terms to represent observables (i.e., variables or measurement concepts). At times (see below), we will use X_C and Y_C to represent *theoretical constructs*, general terms that are not limited to time or place. In Chapter 4 we will discuss theoretical constructs like "falling bodies" and "status" that are drawn from physical and social science respectively.

2. For what we call empirical explanation, Walker (2002) uses the term "historical explanation" because those who develop and use such explanations take a historical approach to the production of knowledge (Berger, Zelditch, and Anderson 1972: *ix*). The ideology that science does not extend beyond observation and generalization from observations is called empiricism. Because we understand that observations and generalizations are not an end, but a step toward theory building and testing, we are not empiricists.

3. R^2 (pronounced "R-squared") is the amount of variation in the dependent measure, *Y*, accounted for or "explained" by variation in the independent measures.

4. More accurately, theories explain statements about relationships between phenomena (see Cohen 1989).

5. A hypothesis or theory can be pronounced "provisionally false" if one or a few isolated observations find negative evidence. Falsification requires replication. (See our discussion of the importance of replication to falsification later in this chapter.)

Notes to Chapter 3

1. Empiricism and empiricist are often used as pejorative terms. We use them in their descriptive sense. An empiricist approach to knowledge "emphasizes the importance of observation and of creating knowledge by amassing observations and generalizing from these observations" (Cohen 1989:16; and see Popper 1962:21ff).

2. While we focus on Mill's method of difference below, the logic of all five canons is consistent with Hempel's (1966) description of methods of discovery.

3. Using the symbols introduced in Chapter 2, *A* and *X* are event-classes. In long form, the symbolic statement can be read, "When an instance of event-class *X* occurs, an instance of event-class *A* precedes it." We use the shortened version for ease of presentation.

4. The *n* weighted average of the standard deviations and the *t* statistic are calculated as follows:

$$\sigma_{(x_1 - x_2)} = \sqrt{\frac{N_1 s_1^2 + N_2 s_2^2}{N_1 + N_2 - 2}} \qquad t = \frac{X_1 - X_2}{\sigma_{(x_1 - x_2)}}$$

5. Modern beer drinkers who marvel that their favorite brew tastes the same in New York or Pocatello can be thankful for Gossett. Guinness was concerned with developing better methods of comparing samples of beer in order to better evaluate and control its quality. To ensure that their competitors' did not learn Gossett's secret for better evaluating samples, Guinness forbade Gossett to publish his method. So, although he knew and corresponded regularly with leading statisticians of his time, Gossett published his work under the pseudonym, Student. Many statistics texts refer to Gossett's method as Student's *t*-test.

6. Chapter 5 discusses artifacts, sources of artifacts (including demand characteristics, subject bias, and experimenter bias), and techniques for minimizing or eliminating artifacts.

7. Many variants of the standard Asch task have been developed. Some use slides to project lines on a screen while others deliver the stimulus materials to computer screens.

8. Those who doubt the plausibility of our alternative interpretation should view the film made of Milgram's experiments. (The film is available through Pennsylvania State University's media sales at http://www.mediasales.psu.edu/.)

9. Orne (1962) used the term "demand characteristic" for those cases in which subjects infer an experimenter's hypotheses, take the role of the "good subject," and behave in ways that tend to confirm the hypotheses. Demand characteristics are discussed in Chapter 5.

10. A number of strategies have been devised for the *repeated* P/D game that make mutual cooperation, and its jointly higher pay, possible. For example, in "tit-for-tat" the player cooperates on the first round and then selects the option selected by the other in the previous round. Two players playing "tit-for-tat" will always cooperate. (See Axelrod 1984.)

11. All research universities and research organizations have institutional review boards (IRBs) that have the responsibility of evaluating and guarding the ethical treatment of subjects. They weigh the benefits that could result from a study in light of the risks to which participants are exposed. Ethics and the role that IRBs have in enforcing ethical standards are discussed in Chapter 5.

12. Simpson's experiments are important exceptions to the need for point-by-point similarities. As we said above, his findings are more easily applied outside the lab because his hybrid design employs theory that has been tested previously and applied outside the lab.

Notes to Chapter 4

1. In Popper's program, *testability* refers to the capacity of a theory to be found false. The more opportunities a theory has of being refuted by evidence, the more testable and more powerful the theory (Popper 1994:88). The history of science has many examples of theories for which no one could initially imagine or devise em-

pirical tests. However, as science progressed many of those theories were tested and evidence found to support—or disconfirm—them.

2. Consider the controversy that surrounds "proof" of cold fusion (Voss 1999). Failure to replicate findings can reflect poor theory, poor design, fraud, or all three. Neither the researchers who reported evidence of cold fusion in 1989 nor others have been able to replicate their results.

3. Survey experiments that test theories are exceptions and are designed by theory.

4. Here M stands for mass and, of course, mass and weight are not the same thing. Mass is the quantity of material in a thing and weight is the force exerted by that mass in a gravitational field. For example, a given mass will have approximately one-third the weight on the moon that it has on earth. Because gravitational force is the same at the two ends of the lever, it cancels out and it is masses that are being compared. It follows that a lever that balances on the earth will balance on the moon as well. Nevertheless, the exposition seems clearer when reference is made to the weights of M and M′.

5. Galileo is said to have used the Tower at Pisa for the experiment.

6. Galileo's experimental apparatus exists today and can be seen in the Science Museum, Florence, Italy. Those interested in the history of science can see central aspects of modern scientific thinking evolve in his work. (See Galileo [1636] 1954.) For simple falling bodies, no diagrams were drawn, but as Galileo's interest turned to projectile motion and other more complex phenomena he drew an array of diagrams. Early diagrams are drawn realistically. For example the analysis of forces on a beam pictured a wooden beam embedded in a rock wall holding up a large weight. Moving to later diagrams realism fades: they become straightforward geometric models

7. Several points must be made here. First, SCT stresses that these criteria are used to identify diffuse characteristics. Second, these are *beliefs* about characteristics. The examples should not be construed as innate properties of individuals or groups. Third, whether a particular attribute like gender is actually a diffuse characteristic is culturally, and therefore, time and place, specific. Females generally have lower status than males in the contemporary United States, but in Sweden males and females are at or near equality. If females and males have equal status in Sweden (or in any other country), gender is not a diffuse characteristic there.

8. Theorists employ a common strategy. In theoretical work, begin with the simplest situation possible. The simplest situations rarely exist in the world of phenomena. Situations in which two people differ on a single status characteristic are almost as rare as frictionless fulcrums and weightless beams.

9. Earnings were announced after each round, peripherals were aware of the inequality, and, as we point out, responded to it.

Notes to Chapter 5

1. The phrase is linked to the Hippocratic Oath but it is not part of the oath. Moreover, neither the intended meaning nor the source of the Latin phrase, "*primum non nocere*" is clear. According to Smith (2005), medical historians have found simi-

lar phrasing in Hippocrates' *Epidemics* (Book I, Chapter XI, [400 BC] 1923). However, it is unlikely that he or Galen, the Roman physician who is another putative author, is the source since both wrote in Hippocrates' native Greek. The phrase's origins are less important than the message. Those whose research involves human subjects must be committed to protecting their health and safety at all costs.

2. The role of the U.S. government in this tragedy is unconscionable. However, we must point out that many medical practitioners and medical organizations, including both the American Medical Association and the predominantly black National Medical Association, were aware of the study and condoned it at least through the late 1960s.

3. For a synopsis of the study see http://www.zimbardo.com. For a dramatization, which may or may not be true to the original, *The Experiment* is available on DVD.

4. Codes of ethics for the American Sociological Association (1999, http://www.asanet.org), the American Anthropological Association (1998, http://www.aaanet.org), and the American Psychological Association (2002, http://www.apa.org) are linked to their respective home pages. These associations update their codes of ethics at periodic intervals.

5. Some research is "exempt" from IRB review, but normally exempt status must be verified by the governing IRB.

6. It is worth noting that Humphreys changed his appearance before he began follow-up interviews with the men he had observed many months earlier. Of course, his subjects did not have that option.

7. It is important to point out that the vast majority of social and behavioral researchers behave ethically. For example, Christina Maslach, a Stanford psychology graduate student, called on researchers to end the Stanford Prison Experiment. Today, Maslach, a professor at California-Berkeley, is a leading expert on the dehumanizing effects of job burnout.

8. We take Ledyard's remarks to mean that subjects' beliefs about the possibility of deception can introduce artifacts into a study. Artifacts affect results and can make them difficult or impossible to interpret. We discuss artifacts, how to reduce or eliminate them, and how to test for them later in this chapter. Here, we focus on how researchers can use good debriefing techniques to reveal deception and to repair its negative effects.

9. Cohen et al. (1996) questioned subjects after debriefing them in the second of their three studies. They report that 89 percent of subjects expressed no anger after they had an opportunity to meet with the insulting confederate. Among the remainder, none reported anger as high as the midpoint on a rating scale that ranged from 0 (not at all angry) to 7 (extremely angry).

10. The investigators also excluded black and Hispanic men to eliminate the possibility of race and ethnic overtones.

11. The situation we describe is very different than Orne's (1962) idea of a pre-experimental inquiry or nonexperiment (Willer 1987). Nonexperiments ask members of a subject pool to behave (i.e., role-play) *as if* they were in an experiment. The scenario we describe is more like improvisational theater without the benefit of minimal

role descriptions. The likelihood of successful coordination in the simple three-actor networks we describe in Chapters 1 and 4 is low. It is substantially lower in the complex, multiple actor structures described in Willer (1999).

12. The reader is reminded that only subjects who administered shocks at the highest level on the shock generator were classified as obedient.

Notes to Chapter 6

1. Toyota Motor Company announced in early summer 2005 that it would build a second assembly plant in Ontario, Canada (Austen 2005). The Canadian and Ontario governments outbid several U.S states.

2. If theories had independent and dependent variables, we could say that variations in the independent variables are introduced so that changes in dependent variables can be tracked. The idea of scope sampling is not new. *A "scope sample" may be defined as a number of natural cases fitting the conditions appropriate to the theory model, which are ranged along the major dimensions of the formal system* (Willer 1967:114). Though more than 40 years old, we know of few studies using the procedure.

References

American Anthropological Association. 1998. *Code of Ethics of the American Anthropological Association.* Washington, DC: American Anthropological Association.

American Psychological Association. 2002. *Ethical Principles of Psychologists and Code of Conduct.* Washington, DC: American Psychological Association.

American Sociological Association. 1999. *Code of Ethics and Policies and Procedures.* Washington, DC: American Sociological Association.

Archimedes. [230 BC] 1897. *The Works of Archimedes.* T. L. Heath (ed.). New York: Dover.

Aristotle. [347 BC] 1962. *Nicomachean Ethics.* Martin Ostwald (tr.). Indianapolis, IN: Bobbs-Merrill.

———. [330 BC] 1961. *Physics.* Richard Hope (tr.). Lincoln: University of Nebraska Press.

Asch, Solomon E. 1958. "Interpersonal Influence." Pp. 174–183 in Eleanor Maccoby, Theodore Newcomb, and Eugene Hartley (eds.), *Readings in Social Psychology,* 3rd Edition. New York: Holt, Rinehart and Winston.

Austen, Ian. 2005. "Toyota Is Said to Be Planning Its Second Factory in Canada." *New York Times,* Late Edition, June 24, 2005.

Axelrod, Robert. 1984. *The Evolution of Cooperation.* New York: Basic Books.

Ayres, Ian. 2006. "Discrimination in Consummated Car Purchases." Chapter 6 in Laura Beth Nielsen and Robert L. Nelson (eds.), *Handbook of Employment Discrimination Research.* New York: Springer.

Bacon, Francis. [1620] 2000. *Novum Organon.* Lisa Jardine and Michael Silverthorne (eds.). Cambridge, UK: Cambridge University Press.

Bales, Robert F. 1950. *Interaction Process Analysis.* Cambridge, MA: Addison-Wesley.

———. 1999. *Social Interaction Systems.* New Brunswick, NJ: Transaction.

Bargh, John A., Mark Chen, and Lara Burrows. 1996. "Automaticity of Social Behavior: Direct Effects of Trait Construct and Stereotype Activation on Action." *Journal of Personality and Social Psychology* 71:230–244.

Bavelas, Alex. 1950. "Communication Patterns in Task-Oriented Groups." *Journal of the Acoustical Society of America* 22:725–730.

Bell, Richard, Henry A. Walker, and David Willer. 2000. "Power, Influence and Legitimacy in Organizations: Implications of Three Theoretical Research Programs." Pp. 131–178 in Samuel B. Bacharach and Edward Lawler (eds.), *Organizational Politics*. Stamford, CT: JAI Press.

Berger, Joseph, Bernard P. Cohen, and Morris Zelditch, Jr. 1966. "Status Characteristics and Expectation States." Pp. 29–46 in Joseph Berger, Morris Zelditch, Jr., and Bo Anderson (eds.), *Sociological Theories in Progress*, vol. 1. Boston: Houghton-Mifflin.

———. 1972. "Status Characteristics and Social Interaction." *American Sociological Review* 37:241–255.

Berger, Joseph and M. Hamit Fisek. 1970. "Consistent and Inconsistent Status Characteristics and the Determination of Power and Prestige Orders." *Sociometry* 33:287–304.

———. 2006. "Diffuse Status Characteristics and the Spread of Status Value: A Formal Theory." *American Journal of Sociology* 111:1038–1079.

Berger, Joseph, M. Hamit Fisek, and Paul Crosbie. 1970. "Multi-characteristic Status Situations and the Determination of Power and Prestige Orders." Technical Report #35. Stanford, CA: Laboratory for Social Research.

Berger, Joseph, M. Hamit Fisek, and Robert Z. Norman. 1998. "The Evolution of Status Expectations: A Theoretical Extension." Pp. 175–205, in Jospeh Berger and Morris Zelditch, Jr. (eds.), *Status, Power and Legitimacy*. New Brunswick, NJ: Transaction.

Berger, Joseph, M. Hamit Fisek, Robert Z. Norman, and Morris Zelditch, Jr. 1977. *Status Characteristics and Social Interaction*. New York: Elsevier.

Berger, Joseph, David Willer, and Morris Zelditch, Jr. 2005. "Theory Programs and Theoretical Problems." *Sociological Theory* 23:127–155.

Berger, Joseph, Morris Zelditch, Jr., and Bo Anderson. 1972. "Historical and Generalizing Orientations in Sociology." Pp. *ix-xxi* in Joseph Berger, Morris Zelditch, Jr., and Bo Anderson (eds.), *Sociological Theories in Progress*, vol. 2. Boston: Houghton-Mifflin.

Berk, Richard A., Kenneth J. Lenihan, and Peter H. Rossi. 1980. "Crime and Poverty: Some Experimental Evidence from Ex-Offenders." *American Sociological Review* 45:766–786.

Brennan, James. 1981. "Some Experimental Structures." Pp. 198–206 in David Willer and Bo Anderson (eds.), *Networks, Exchange and Coercion*. New York: Elsevier.

Burgess, Ernest. 1929. "Basic Social Data." Pp. 47–66 in T. V. Smith and L. D. White. *Chicago: An Experiment in Social Science Research*. Chicago: University of Chicago Press.

Campbell, Donald T. and Julian C. Stanley. 1966. *Experimental and Quasi-Experimental Designs for Research*. Chicago: Rand McNally.

Cohen, Bernard P. 1989. *Developing Sociological Knowledge*, 2nd Edition. Chicago: Nelson-Hall.

Cohen, Dov, Richard E. Nisbett, Brian F. Bowdle, and Norbert Schwarz. 1996. "Insult, Aggression and the Southern Culture of Honor: An 'Experimental Ethnography.'" *Journal of Personality and Social Psychology* 70:945–960.

Cohen, Elizabeth G. 1998. "Complex Instruction." *European Journal of Intercultural Studies* 9:127–131.

Cohen, Elizabeth G., Rachel A. Lotan, Beth A. Scarloss, and Adele R. Arellano. 1999. "Complex Instruction: Equity in Cooperative Learning Classrooms." *Theory into Practice* 38:80–86.

Corra, Mamadi. 2005. "Separation and Exclusion: Distinctly Modern Conceptions of Power?" *Canadian Journal of Sociology* 30:41–70.

Corra, Mamadi and David Willer. 2002. "The Gatekeeper." *Sociological Theory* 20:180–207.

Dunn, Cynthia M. and Gary Chadwick. 2001. *Protecting Study Volunteers in Research: A Manual for Investigative Sites*. Boston: CenterWatch.

Einstein, Albert. [1934] 1954. "On the Method of Theoretical Physics." Pp.270–276 in Carl Seelig (ed.), *Ideas and Opinions*. New York: Crown.

Fermat, Pierre. [1662] 1896. "Letter to Cureau de la Chambre." Pp. 457–463 in Paul Tannery and Charles Henry (eds.), *Oeuvres de Fermat*, vol. 2. Paris: Gauthier-Villars et Fils.

Fisher, Ronald A. 1935. *The Design of Experiments*. London: Oliver and Boyd.

———. 1956. *Statistical Methods and Scientific Inference*. Edinburgh: Oliver and Boyd.

Flament, Claude. 1962. *Applications of Graph Theory to Group Structure*. Englewood Cliffs, NJ: Prentice-Hall.

Freese, Lee and Jane Sell. 1980. "Constructing Axiomatic Theories in Sociology." Pp. 263–368 in Lee Freese (ed.), *Theoretical Methods in Sociology: Seven Essays*. Pittsburgh: University of Pittsburgh Press.

Galilei, Galileo. [1636] 1954. *Dialogues Concerning Two New Sciences*. Henry Crew and Alfonso DeSalvio (tr.). New York: Dover.

Geller, Daniel M. 1982. "Alternatives to Deception: Why, What, and How?" Chapter 2, pp. 39–55 in Joan E. Sieber (ed.), *The Ethics of Social Research: Surveys and Experiments*. New York: Springer-Verlag.

Haney, Craig, Curtis Banks, and Philip G. Zimbardo. 1973. "Interpersonal Dynamics in a Simulated Prison." *International Journal of Criminology and Penology* 1:69–97.

Harary, Frank, Robert Z. Norman, and Dorwin Cartwright. 1965. *Structural Models: An Introduction to the Theory of Directed Graphs*. New York: John Wiley and Sons.

Heath, Thomas L. 1897. "Archimedes." Pp. *xv–xxiii* in Thomas L. Heath (ed.), *The Works of Archimedes*. New York: Dover.

Hempel, Carl G. 1966. *Philosophy of Natural Science*. Englewood Cliffs, NJ: Prentice-Hall.

Hippocrates. [400 BC] 1923. *Ancient Medicine; Airs; Waters; Places; Epidemics 1 & 3; The Oath; Precepts; Nutriment*. W. H. S. Jones (tr.). Cambridge, MA: Harvard University Press.

Holmes, David S. 1976a. "Debriefing After Psychological Experiments: I. Effectiveness of Postdeception Dehoaxing." *American Psychologist* 31:858–867.

———. 1976b. "Debriefing After Psychological Experiments: I. Effectiveness of Postdeception Desensitizing." *American Psychologist* 31:868–875.

Hughes, Everett C. 1945. "Dilemmas and Contradictions of Status." *American Journal of Sociology* 50:353–359.

Humphreys, Laud. 1970. *Tearoom Trade: Impersonal Sex in Public Places*. Chicago: Aldine.

Jones, James H. 1981. *Bad Blood: The Tuskegee Syphilis Experiment*. New York: Free Press.

Kelman, Herbert C. 1967. "Human Use of Human Subjects: The Problem of Deception in Social Psychological Experiments." *Psychological Bulletin* 67:1–11.

Kruglanski, Arie. 1975. "The Human Subject in the Psychology Experiment: Fact and Artifact." Pp. 101–147 in Leonard Berkowitz (ed.), *Advances in Experimental Social Psychology*, Volume 8. New York: Academic Press.

Kuwabara, Ko. 2005. "Nothing to Fear But Fear Itself?: Fear of Fear, Fear of Greed and Gender Effects in Two-Person Asymmetric Social Dilemmas." *Social Forces* 84:1257–1272.

Leahey, Erin. 2005. "Alphas and Asterisks: The Development of Statistical Significance Testing Standards in Sociology." *Social Forces* 84:1–24.

Lederman, Leon. 1993. *The God Particle: If the Universe Is the Answer, What Is the Question*? With Dick Teresi. New York: Delta.

Ledyard, John. 1995. "Public Goods: A Survey of Experimental Research." Pp. 111–194 in John Kagel and Alvin Roth (eds.), *The Handbook of Experimental Economics*. Princeton, NJ: Princeton University Press.

Lewin, Kurt, Ronald Lippitt, and Ralph K. White. 1939. "Patterns of Aggressive Behavior in Experimentally Created 'Social Climates'." *Journal of Social Psychology* 10:271–299.

Lichter, Daniel T., Diane K. McLaughlin, and David C. Ribar. 1997. "Welfare and the Rise in Female-Headed Families." *American Journal of Sociology* 103:112–143.

Lieberson, Stanley. 1985. *Making It Count*. Berkeley: University of California Press.

Lieberson, Stanley and Freda Lynn. 2002. "Barking up the Wrong Branch: Scientific Alternatives to the Current Model of Sociological Science." *Annual Review of Sociology* 28:1–19.

Lucas, Jeffery W. 2003. "Theory-Testing, Generalization and the Problem of External Validity." *Sociology Theory* 21:236–253.

Lucas, Jeffrey W., Corina Graif, and Michael J. Lovaglia. 2006. "Misconduct in the Prosecution of Severe Crimes: Theory and Experimental Test." *Social Psychology Quarterly* 69:97–107.

McNeill, Paul. 1993. *The Ethics and Politics of Human Experimentation.* Cambridge, UK: Cambridge University Press.

———. 1997. "Paying People to Participate in Research: Why Not?" *Bioethics* 11:390–396.

Mann, Michael. 1986. *The Sources of Social Power.* New York: Cambridge University Press.

Mantell, David M. 1971. "The Potential for Violence in Germany." *Journal of Social Issues* 27:101–112.

Marx, Karl. [1857] 1973. *Grundrisse.* New York: Vintage.

———. [1867] 1967. *Capital.* New York: International Publishers.

Milgram, Stanley. 1965. "Some Conditions of Obedience and Disobedience to Authority." *Human Relations* 18:57–75.

———. 1974. *Obedience to Authority.* New York: Harper & Row.

Mill, John S. [1843] 1967. *A System of Logic.* London: Longmans, Green.

Moore, James C. 1968. "Status and Influence in Small Group Interactions." *Sociometry* 31:47–63.

Nagel, Ernest. 1961. *The Structure of Science.* New York: Harcourt Brace and World.

Nagel, Jack. 1975. *The Descriptive Analysis of Power.* New Haven, CT: Yale University Press.

National Institutes of Health. 1979. *The Belmont Report: Ethical Principles and Guidelines for the Protection of Human Subjects of Research.* Washington, DC: Department of Health, Education and Welfare.

Newton, Isaac. [1686] 1966. *Principia Mathematica.* A. Motte (tr.). Berkeley: University of California Press.

Orne, Martin T. 1962. "On the Social Psychology of the Psychological Experiment: With Particular Reference to Demand Characteristics and their Implications." *American Psychologist* 17:776–783.

———. 1969. "Demand Characteristics and the Concept of Quasi-Controls." In Robert Rosenthal and Ralph L. Rosnow (eds.), *Artifact in Behavioral Research.* New York: Academic Press.

Piliavin, Irving M., Judith Rodin, and Jane A. Piliavin. 1969. "Good Samaritanism: An Underground Phenomenon." *Journal of Personality and Social Psychology* 13:289–299.

Popper, Karl R. 1962. *Conjectures and Refutation.* New York: Basic Books.

———. 1994. *The Myth of Framework.* London: Routledge.

Prechel, Harland. 2000. *Big Business and the State.* Albany, NY: State University of New York Press.

Rosenholtz, Susan J. and Elizabeth G. Cohen. 1983. "Back to Basics and the Desegregated School." *Elementary School Journal* 83:515–527.

Rosenthal, Robert. 1966. *Experimenter Effects in Behavioral Research.* New York: Appleton-Century-Crofts.

Rosenthal, Robert and Lenore Jacobson. 1968. *Pygmalion in the Classroom.* New York: Holt, Rinehart and Winston.

Schachter, Stanley and Jerome E. Singer. 1962. "Cognitive, Social, and Physiological Determinants of Emotional State." *Psychological Review* 69:379–399.

Selltiz, Claire, Marie Jahoda, Morton Deutsch, and Stuart W. Cook. 1959. *Research Methods in Social Relations,* Revised Edition. New York: Holt, Rinehart and Winston.

Sieber, Joan E. 1992. *Planning Ethically Responsible Research: A Guide for Students and Internal Review Boards.* Newbury Park, CA: Sage.

Simpson, Brent. 2003. "Sex, Fear, and Greed." *Social Forces* 82:35–52.

Skvoretz, John and David Willer. 1991. "Power in Exchange Networks: Setting and Structural Variations." *Social Psychology Quarterly* 54:224–238.

Smith, Cedric M. 2005. "Origin and Uses of *Primum Non Nocere*—Above All, Do No Harm!" *Journal of Clinical Pharmacology* 45:371–377.

Strodtbeck, Fred L., Rita M. James, and Charles Hawkins. 1957. "Social Status in Jury Deliberations." *American Sociological Review* 22:713–719.

Stouffer, Samuel A., Edward A. Suchman, Leland C. Devinney, Shirley A. Star, and Robin M. Williams, Jr. 1949. *The American Soldier: Studies in Social Psychology in World War II.* Princeton, NJ: Princeton University Press.

Thye, Shane R., David Willer, and Barry Markovsky. 2006. "From Status to Power: New Models at the Intersection of Two Theories." *Social Forces* 84:1471–1495.

Voss, David. 1999. "What Ever Happened to Cold Fusion?" *Physics World* (http://physicsweb.org/articles/world/12/3/8/1 [accessed 4 December 2003]).

Wagner, David and Joseph Berger. 1985. "Do Sociological Theories Grow?" *American Journal of Sociology* 90:697–728.

Walker, Henry A. 2002. "Three Faces of Explanation: A Strategy for Building Cumulative Knowledge." Pp. 15–31 in Jacek Szmatka, Michael Lovaglia, and Kinga Wysienska (eds.), *The Growth of Social Knowledge.* Westport, CT: Praeger.

Walker, Henry A., Shane R. Thye, Brent Simpson, Michael Lovaglia, David Willer, and Barry Markovsky. 2000. "Network Exchange Theory: Recent Developments and New Directions." *Social Psychology Quarterly* 63: 324–337.

Walker, Henry A. and Morris Zelditch, Jr. 1993. "Power, Legitimation, and the Stability of Authority: A Theoretical Research Program." Pp. 364–381 in Joseph Berger and Morris Zelditch, Jr. (eds.), *Theoretical Research Programs: Studies in the Growth of Theory.* Stanford, CA: Stanford University Press.

Weber, Max. [1896] 1976. "The Social Causes of the Decay of Ancient Civilization." In Russell Kahl (ed.), *Studies in Explanation.* R. Frank (tr.). Englewood Cliffs, NJ: Prentice-Hall.

———. [1911] 1951. *The Religion of China.* New York: Free Press.

———. [1918] 1968. *Economy and Society.* Berkeley: University of California Press.

———. [1904] 1958. *The Protestant Ethic and the Spirit of Capitalism.* Talcott Parsons (tr.). With a foreword by R. H. Tawney. New York: Scribner.

Weber, Stephen J. and Thomas D. Cook. 1972. "Subject Effects in Laboratory Research: An Examination of Subject Roles, Demand Characteristics, and Valid Inference." *Psychological Bulletin* 77(4):273–295.

Wilkinson, Martin and Andrew Moore. 1997. "Inducement in Research." *Bioethics* 11:373–389.

Willer, David. 1967. *Scientific Sociology.* Englewood Cliffs, NJ: Prentice-Hall.

———. 1984. "Analysis and Composition as Theoretic Procedures." *The Journal of Mathematical Sociology* 10:241–270.

———. 1987. *Theory and the Experimental Investigation of Social Structures.* New York: Gordon and Breach.

———. 1999. *Network Exchange Theory.* Westport, CT: Praeger.

———. 2003. "Power-at-a-Distance." *Social Forces* 81:1295–1334.

Willer, David and Bo Anderson (eds.). 1981. *Networks Exchange and Coercion: The Elementary Theory and its Applications.* New York: Elsevier.

Willer, David, Brent Simpson, Jacek Szmatka, and Joanna Mazur. 1997. "Social Theory and Historical Explanation." *Humboldt Journal of Social Relations* 22:63–84.

Willer, David and Murray Webster, Jr. 1970. "Theoretical Concepts and Observables." *American Sociological Review* 35:748–757.

Zelditch, Morris, Jr. 1992. "Interpersonal Power." Pp. 994–1001 in Edgar F. Borgatta and Marie L. Borgatta (eds.), *Encyclopedia of Sociology.* New York: Macmillan.

Zimbardo, Philip G., Christina Maslach, Craig Haney. 2000. "Reflections on the Stanford Prison Experiment: Genesis, Transformations, Consequences." Chapter 11, pp. 193–237 in Thomas Blass (ed.), *Obedience to Authority: Current Perspectives on the Milgram Paradigm.* Mahwah, NJ: Erlbaum.

Index